Multivariate Statistics

THEORY AND APPLICATIONS

Proceedings of IX Tartu Conference on Multivariate Statistics and XX International Workshop on Matrices and Statistics

Multivariate Statistics

THEORY AND APPLICATIONS

Tartu, Estonia, 26 June – 1 July 2011

Editor

Tõnu Kollo

University of Tartu, Estonia

World Scientific

NEW JERSEY • LONDON • SINGAPORE • BEIJING • SHANGHAI • HONG KONG • TAIPEI • CHENNAI

Published by

World Scientific Publishing Co. Pte. Ltd.
5 Toh Tuck Link, Singapore 596224
USA office: 27 Warren Street, Suite 401-402, Hackensack, NJ 07601
UK office: 57 Shelton Street, Covent Garden, London WC2H 9HE

British Library Cataloguing-in-Publication Data
A catalogue record for this book is available from the British Library.

MULTIVARIATE STATISTICS: THEORY AND APPLICATIONS
Proceedings of IX Tartu Conference on Multivariate Statistics and
XX International Workshop on Matrices and Statistics

ISBN 978-981-4449-39-7

Printed in Singapore.

PREFACE

This volume consists of selected papers presented at the IX Tartu Conference on Multivariate Statistics organized jointly with the XX International Workshop on Matrices and Statistics. The conference was held in Tartu, Estonia from 26 June to 1 July 2011. More than 100 participants from 30 countries presented in four days recent devolopments on various topics of multivariate statistics.

The papers cover wide range of problems in modern multivariate statistics including distribution theory and estimation, different models of multivariate analysis, design of experiments, new developments in high-dimensional statistics, sample survey methods, graphical models and applications in different areas: medicine, transport, life and social sciences.

The Keynote Lecture by Professor N. Balakrishnan was delivered as the Samuel Kotz Memorial Lecture. Thorough treatment of multivariate exponential dispersion models is given by Professor B. Jørgensen. A new general approach to sampling plans is suggested by Professor Y. K. Belyaev. Professor E. Ahmed compares different strategies of estimating regression parameters. C. M. Cuadras introduces a generalization of Farley-Gumbel-Morgenstern distributions.

As Editor I am thankful to the authors who have presented interesting and valuable results for publishing in the current issue. The book will be useful for researchers and graduate students who work in multivariate statistics. The same time numerous applications can give useful ideas to scientists in different areas of research. My special thanks go to the anonymous Referees who have done great job and spent lot of time with reading the papers. Due to their comments and suggestions the presentation of the material has been improved and the quality of the papers has risen. I am extremely thankful to the technical secretary of the volume Dr. Ants Kaasik who has efficiently organised correspondence with the authors and Referees.

Tõnu Kollo
Editor

Tartu, Estonia
October 2012

ORGANIZING COMMITTEES

PROGRAMME COMMITTEE

of The 9th Tartu Conference on Multivariate Statistics and The 20th International Workshop on Matrices & Statistics

D. von Rosen (Chairman)	– Swedish University of Agricultural Sciences, Linköping University, Sweden
G. P. H. Styan (Honorary Chairman of IWMS)	– McGill University, Canada
T. Kollo (Vice-Chairman)	– University of Tartu, Estonia
S. E. Ahmed	– University of Windsor, Canada
J. J. Hunter	– Auckland University of Technology, New Zealand
S. Puntanen	– University of Tampere, Finland
G. Trenkler	– Technical University of Dortmund, Germany
H. J. Werner	– University of Bonn, Germany

ORGANIZING COMMITTEE

of The 9th Tartu Conference on Multivariate Statistics and The 20th International Workshop on Matrices & Statistics

K. Pärna (Chairman)	– University of Tartu, Estonia
A. Kaasik (Conference Secretary)	– University of Tartu, Estonia

CONTENTS

VARIABLE SELECTION AND POST-ESTIMATION OF REGRESSION PARAMETERS USING QUASI-LIKELIHOOD APPROACH

S. FALLAHPOUR

Department of Mathematics and Statistics, University of Windsor,
Windsor, ON N9B 3P4, Canada
E-mail: fallahp@uwindsor.ca
www.uwindsor.ca

S. E. AHMED

Department of Mathematics, Brock University,
St. Catharines, ON L2S 3A1, Canada
E-mail: sahmed5@brocku.ca

In this paper, we suggest the pretest estimation strategy for variable selection and estimating the regression parameters using quasi-likelihood method when uncertain prior information (UPI) exist. We also apply the lasso-type estimation and variable selection strategy and compare the relative performance of lasso with the pretest and quasi-likelihood estimators. The performance of each estimator is evaluated in terms of the simulated mean square error. Further, we develop the asymptotic properties of pretest estimator (PTE) using the notion of asymptotical distributional risk, and compare it with the unrestricted quasi-likelihood estimator (UE) and restricted quasi-likelihood estimator (RE), respectively. The asymptotic result demonstrates the superiority of pretest strategy over the UE and RE in meaningful part of the parameter space. The simulation results show that when UPI is correctly specified the PTE outperforms lasso.

Keywords: Pretest Estimator; Quasi-likelihood; Asymptotic Distributional Bias and Risk; Lasso.

1. Introduction

First, we present the quasi-likelihood (QL) function and describe its properties. The term quasi-likelihood was introduced by Robert Wedderburn[1] to describe a function which has similar properties to the log-likelihood function, except that a QL function is not the log-likelihood corresponding to any actual probability distribution. Instead of specifying a probability distribution for the data, only a relationship between the mean and the variance is specified in the form of a variance function when given the variance as a function of the mean. Thus, QL is based on the assumption of only the first two moments of the response variable.

Consider the uncorrelated data y_i with $E(y_i) = \mu_i$ and $var(y_i) = \phi V(\mu_i)$, where μ_i is to be modeled in terms of a p-vector of parameters $\boldsymbol{\beta}$, the variance function $V(.)$ is assumed a known function of μ_i, and ϕ is a multiplicative factor known as the dispersion parameter or scale parameter that is estimated from the data. Suppose that for each observation y_i, the QL function $Q(y_i; \mu_i)$ is given by

$$Q(y_i; \mu_i) = \int_{y_i}^{\mu_i} \frac{y_i - t}{\phi V(t)} dt.$$

For each observation, the quasi-score function $U(y_i, \mu_i)$ is defined by the relation

$$U(y_i, \mu_i) = \frac{\partial Q}{\partial \mu_i} = \frac{y_i - \mu_i}{\phi V(\mu_i)}.$$

Let us consider n independent observations $\mathbf{y} = (y_1, y_2, \ldots, y_n)'$ with a set of predictor values $\mathbf{x}_i = (x_{i1}, x_{i2}, \ldots, x_{ip})'$. In the generalized linear form we have

$$E(y_i) = \mu_i, \qquad g(\mu_i) = \sum_{r=1}^{p} \beta_r x_{ir} \qquad i = 1, \ldots, n,$$

with the generalized form of variance

$$var(y_i) = \phi V(\mu_i) \qquad i = 1, \ldots, n,$$

where $g(.)$ is the link function which connects the random component $\mathbf{y}$ to the systematic components $\mathbf{x}_1, \mathbf{x}_2, \ldots, \mathbf{x}_p$. It is obvious that μ_i is a function of $\boldsymbol{\beta}$ since $\mu_i = g^{-1}(\mathbf{x}_i'\boldsymbol{\beta})$, so we can rewrite $\boldsymbol{\mu} = \boldsymbol{\mu}(\boldsymbol{\beta})$.

The statistical objective is to estimate the regression parameters $\beta_1, \beta_2, \ldots, \beta_p$. Since the observations are independent by assumption, the QL for the complete data is the sum of the individual quasi-likelihoods:

$$Q(\mathbf{y}, \boldsymbol{\mu}) = \sum_{i=1}^{n} Q(y_i, \mu_i).$$

The estimation of the regression parameters $\boldsymbol{\beta}$ is obtained by differentiating $Q(\mathbf{y}, \boldsymbol{\mu})$ with respect to $\boldsymbol{\beta}$, which may be written in the form of $\mathbf{U}(\hat{\boldsymbol{\beta}}) = 0$, where

$$\mathbf{U}(\boldsymbol{\beta}) = \mathbf{D}'\mathbf{V}^{-1}(\boldsymbol{\mu})(\mathbf{y} - \boldsymbol{\mu})/\phi$$

is called the quasi-score function and $\hat{\boldsymbol{\beta}}$ is the unrestricted maximum quasi-likelihood estimator (UE) of $\boldsymbol{\beta}$. Here, $\mathbf{D}$ is a $n \times p$ matrix and the components

$$D_{ir} = \frac{\partial \mu_i}{\partial \beta_r} \qquad i = 1, 2, \ldots, n \qquad r = 1, 2, \ldots, p$$

are the derivatives of $\boldsymbol{\mu}(\boldsymbol{\beta})$ with respect to the parameters. Since the data are independent, $\mathbf{V}(\hat{\boldsymbol{\mu}})$ can be considered in the form of a diagonal matrix $\mathbf{V}(\boldsymbol{\mu}) = diag\{V_1(\mu_1), \ldots, V_n(\mu_n)\}$, where $V_i(\mu_i)$ is a known function depending only on the ith component of the mean vector $\boldsymbol{\mu}$. Wedderburn[1] and McCullagh[2] show that quasi likelihoods and their corresponding maximum quasi-likelihood estimates have many properties similar to those of likelihoods and their corresponding maximum likelihood estimates.

McCullagh[2] showed that, under certain regularity conditions, the UE ($\hat{\boldsymbol{\beta}}$) is consistent estimator of $\boldsymbol{\beta}$, and

$$\sqrt{n}(\hat{\boldsymbol{\beta}} - \boldsymbol{\beta}) \sim N_p(\mathbf{0}, \phi\, \boldsymbol{\Sigma}^{-1}),$$

where $\boldsymbol{\Sigma} = \left(\mathbf{D}'\mathbf{V}^{-1}(\boldsymbol{\mu})\mathbf{D}\right)$. The covariance matrix $\boldsymbol{\Sigma}$ and ϕ can be estimated using $\hat{\boldsymbol{\beta}}$

$$\hat{\boldsymbol{\Sigma}} = \hat{\mathbf{D}}'\mathbf{V}^{-1}(\hat{\boldsymbol{\mu}})\hat{\mathbf{D}}, \qquad \hat{\phi} = \frac{1}{n-p}\sum_i (y_i - \hat{\mu}_i)^2 / V_i(\hat{\mu}_i),$$

where $\hat{\mu}_i = \mu_i(\hat{\boldsymbol{\beta}})$.

The rest of this paper is organized as follows. In Section 2, we suggest pretest estimation strategy and lasso or absolute penalty estimator (APE). Section 3 provides and compares the asymptotic results of the estimators. In Section 4, we demonstrate via simulation that the suggested strategies have good finite sample properties. Section 5 offers concluding remarks.

2. Improved Estimation and Variable Selection Strategies

2.1. *Pretest Estimations*

In this section we consider the estimation problem for the QL models when some prior information (non-sample information (NSI) or uncertain prior information (UPI)) on parameters $\boldsymbol{\beta}$ is available. The prior information about the subset of $\boldsymbol{\beta}$ can be written in terms of a restriction and we are interested in establishing estimation strategy for the parameters when they are subject to constraint

$$\mathbf{F}'\boldsymbol{\beta} = \mathbf{d},$$

where $\mathbf{F}$ is a $p \times q$ full rank matrix with rank $q \leq p$ and $\mathbf{d}$ is a given $q \times 1$ vector of constants. Under the restriction, it is possible to obtain the estimators of the parameters of the the sub-model, commonly known as the restricted maximum quasi-likelihood estimator or simply restricted estimator (RE). Indeed, following Heyde,[3] the RE can be written as

$$\tilde{\boldsymbol{\beta}} = \hat{\boldsymbol{\beta}} - \boldsymbol{\Sigma}^{-1}\mathbf{F}(\mathbf{F}'\boldsymbol{\Sigma}^{-1}\mathbf{F})^{-1}(\mathbf{F}'\hat{\boldsymbol{\beta}} - \mathbf{d}).$$

Generally speaking, $\tilde{\boldsymbol{\beta}}$ performs better than $\hat{\boldsymbol{\beta}}$ when UPI is true. On the other hand, if the UPI is not correct, the $\tilde{\boldsymbol{\beta}}$ may be considerably biased and inefficient. We refer to Ahmed[4] and Ahmed et al.[5] for some discussions on this point.

In an effort to obtain a compromised estimator of $\boldsymbol{\beta}$ we implement the pretest estimation strategy. The pretest estimator (PTE) based on the UE and RE is defined as

$$\hat{\boldsymbol{\beta}}^{PT} = \hat{\boldsymbol{\beta}} - (\hat{\boldsymbol{\beta}} - \tilde{\boldsymbol{\beta}})I(D_n < c_{q,\alpha}),$$

where D_n is the test-statistic to test the null-hypothesis $H_0 : \mathbf{F}'\boldsymbol{\beta} - \mathbf{d} = \mathbf{0}$ defined in Proposition 3.1, $c_{q,\alpha}$ is the upper α-level critical value of the χ^2 distribution with q degrees of freedom, and $I(A)$ is the indicator function of the set A. Thus, $\hat{\boldsymbol{\beta}}^{PT}$ chooses $\tilde{\boldsymbol{\beta}}$ when H_0 is tenable, otherwise $\hat{\boldsymbol{\beta}}$. For some useful discussions on pretest estimation strategy, we refer to Ahmed and Liu[6] among others.

In Section 3, we derive the asymptotic properties of these estimators including asymptotic distributional bias (ADB) and asymptotic distributional risk (ADR) and we show that the PTE is superior to the RE and UE.

2.2. *Absolute Penalty Estimator (APE)*

The absolute penalty estimation or L_1 penalization approach is used for both estimating and shrinking the model parameters, and has become a very popular technique among researchers. APE/Lasso is a regularization technique for simultaneous estimation and variable selection. APE/lasso coefficients are the solutions to the L_1 optimization problem

$$\hat{\boldsymbol{\beta}}_{lasso} = \underset{\boldsymbol{\beta}}{argmin}\Big\{\sum_{i=1}^{n}(y_i - \mathbf{x}_i'\boldsymbol{\beta})^2 + \lambda\sum_{j=1}^{p}|\beta_j|\Big\},$$

where λ is the tuning parameter. This method has become a popular model selection procedure since it shrinks some coefficients and, because of its

L_1 penalty, the method will set many of the coefficients exactly equal to zero. When λ is large enough, the constraint has no effect; however, for smaller values of $\lambda(\lambda \geq 0)$ the solutions are shrunken versions of the least square estimates, and often some of them are equal to zero. In recent years, there has been an enormous amount of research devoted to regularization methods for different models.

Efron et al.[7] developed an efficient algorithm for computing the lasso for linear regression models. Park and Hastie[8] proposed an L_1 regularization procedure for fitting generalized linear models. It is similar to the lasso procedure, in which the loss function is replaced by the negative log-likelihood of any distribution in the exponential family; i.e.,

$$\hat{\boldsymbol{\beta}} = \underset{\boldsymbol{\beta}}{argmin}\Big\{ -\ell(\boldsymbol{\beta}) + \lambda \sum_{j=1}^{p} |\beta_j| \Big\},$$

where $\ell(\boldsymbol{\beta})$ is the log-likelihood of the underlying GLM. Ahmed et al.[9] and Fallahpour et al.[10] developed shrinkage and variable selection method for partially linear models with uncorrelated and correlated errors, respectively. For a review on other available techniques, we refer to Friedman et al.[11] and references therein.

In this paper, we apply this strategy for estimating regression parameters using a quasi-likelihood approach. We first generated observations from the quasi-Poisson model where $E(y) = \mu$ and $Var(y) = \theta\mu$ where θ is the dispersion parameter and then to obtain the model parameter estimates based on L_1 penalty, suggested by Trevor Hastie in our communication with him, we used the glment-package[12] in R software[13] . The results are shown in Section 4.

3. Asymptotic Distribution Bias and Risk

In this section, we derive the asymptotic properties of the estimators. For this aim, we consider a sequence of local alternatives $\{K_n\}$ given by

$$K_n : \mathbf{F}'\boldsymbol{\beta} = \mathbf{d} + \frac{\boldsymbol{\omega}}{\sqrt{n}},$$

where $\boldsymbol{\omega}$ is a fixed q-column vector. To study the asymptotic quadratic bias and risks of $\hat{\boldsymbol{\beta}}, \tilde{\boldsymbol{\beta}}$ and $\hat{\boldsymbol{\beta}}^{PT}$ a quadratic loss function using a positive definite matrix (p.d.m) $\mathbf{M}$ is defined;

$$L(\boldsymbol{\beta}^{*}, \boldsymbol{\beta}) = n(\boldsymbol{\beta}^{\star} - \boldsymbol{\beta})'\mathbf{M}(\boldsymbol{\beta}^{\star} - \boldsymbol{\beta}),$$

where $\boldsymbol{\beta}^\star$ can be any estimator of $\hat{\boldsymbol{\beta}}, \tilde{\boldsymbol{\beta}}$ and $\hat{\boldsymbol{\beta}}^{PT}$. Suppose for the estimator $\boldsymbol{\beta}^*$ of $\boldsymbol{\beta}$ the asymptotic distribution function exists and can be denoted by

$$F(\mathbf{x}) = \lim_{n\to\infty} P(\sqrt{n}(\boldsymbol{\beta}^\star - \boldsymbol{\beta}) \leq \mathbf{x}|K_n)$$

where $F(\mathbf{x})$ is nondegenerate. The performance of the estimators can be evaluated by comparing their asymptotic distributional risk (ADR). An estimator with a smaller ADR is preferred. The asymptotic distributional risk (ADR) of $\boldsymbol{\beta}^*$ is

$$R(\boldsymbol{\beta}^*, \mathbf{M}) = tr\left(\mathbf{M}\int_{\mathbb{R}^p}\int \mathbf{x}\mathbf{x}'dF(\mathbf{x})\right) = tr(\mathbf{M}\boldsymbol{\Gamma}),$$

where $\boldsymbol{\Gamma}$ is the asymptotic covariance of the estimators, defined as

$$\boldsymbol{\Gamma}(\boldsymbol{\beta}^\star) = E\lim_{n\to\infty}\left[n(\boldsymbol{\beta}^\star - \boldsymbol{\beta})(\boldsymbol{\beta}^\star - \boldsymbol{\beta})'\right].$$

Furthermore we present the asymptotic distributional bias (ADB) of the proposed estimators as

$$ADB(\boldsymbol{\beta}^\star) = E\lim_{n\to\infty}(\sqrt{n}(\boldsymbol{\beta}^\star - \boldsymbol{\beta})).$$

Note that under fixed alternatives all the estimators are equivalent to $\hat{\boldsymbol{\beta}}$, whereas $\tilde{\boldsymbol{\beta}}$ has an unbounded risk. For this reason, to obtain the non-degenerate asymptotic distribution, we consider the local alternatives. For establishing the asymptotic results of the proposed estimators, we need the following regularity conditions:

(1) Weak conditions on the third derivative of $E(\mathbf{y}) = \boldsymbol{\mu}(\boldsymbol{\beta})$ and the third moments of $\mathbf{y}$ are finite.

(2) $\lim_{n\to\infty} n^{-1}\mathbf{D}_n'\mathbf{V}^{-1}(\mu)\mathbf{D}_n = \boldsymbol{\Sigma}$, finite and positive definite matrix.

Under the above regularity conditions, the UE and RE of $\boldsymbol{\beta}$ are consistent and they are asymptotically normal under the local alternative. Now let:

$$\xi_n = \sqrt{n}(\hat{\boldsymbol{\beta}} - \tilde{\boldsymbol{\beta}}), \qquad \zeta_n = \sqrt{n}(\tilde{\boldsymbol{\beta}} - \boldsymbol{\beta}), \qquad \varrho_n = \sqrt{n}(\hat{\boldsymbol{\beta}} - \boldsymbol{\beta})$$

$$\boldsymbol{\Sigma}^* = \boldsymbol{\Sigma}^{-1} - \boldsymbol{\Sigma}^{-1}\mathbf{F}(\mathbf{F}'\boldsymbol{\Sigma}^{-1}\mathbf{F})^{-1}\mathbf{F}'\boldsymbol{\Sigma}^{-1}, \qquad \boldsymbol{\delta} = \boldsymbol{\Sigma}^{-1}\mathbf{F}(\mathbf{F}'\boldsymbol{\Sigma}^{-1}\mathbf{F})^{-1}\boldsymbol{\omega},$$

the joint distribution of the UE and RE are derived below.

Proposition 3.1. *If (1) and (2) hold; then under local alternative K_n, as $n \to \infty$ we have*

$$\begin{pmatrix} \varrho_n \\ \xi_n \end{pmatrix} \sim \mathcal{N}_{2p}\left\{ \begin{pmatrix} \mathbf{0} \\ \boldsymbol{\delta} \end{pmatrix}, \phi \begin{pmatrix} \boldsymbol{\Sigma}^{-1} & \boldsymbol{\Sigma}^{-1} - \boldsymbol{\Sigma}^{*} \\ \boldsymbol{\Sigma}^{-1} - \boldsymbol{\Sigma}^{*} & \boldsymbol{\Sigma}^{-1} - \boldsymbol{\Sigma}^{*} \end{pmatrix} \right\}$$

$$\begin{pmatrix} \zeta_n \\ \xi_n \end{pmatrix} \sim \mathcal{N}_{2p}\left\{ \begin{pmatrix} -\boldsymbol{\delta} \\ \boldsymbol{\delta} \end{pmatrix}, \phi \begin{pmatrix} \boldsymbol{\Sigma}^{*} & 0 \\ 0 & \boldsymbol{\Sigma}^{-1} - \boldsymbol{\Sigma}^{*} \end{pmatrix} \right\}.$$

Proof. See Proofs. □

Corollary 3.1. *If Proposition 3.1 holds, then $\phi^{-1}\xi_n'\mathbf{F}(\mathbf{F}'\Sigma^{-1}\mathbf{F})^{-1}\mathbf{F}'\xi_n \xrightarrow[n\to\infty]{D} \chi_q^2(\Delta)$, where $\chi_q^2(\Delta)$ is a non-central chi-square distribution with q degrees of freedom and non-centrality parameter*
$\Delta = \phi^{-1}\boldsymbol{\omega}'(\mathbf{F}'\Sigma^{-1}\mathbf{F})^{-1}\boldsymbol{\omega}$.

Note that the covariance matrix Σ and ϕ can be estimated using $\hat{\boldsymbol{\beta}}$

$$\hat{\Sigma} = \hat{\mathbf{D}}'\mathbf{V}^{-1}(\hat{\boldsymbol{\mu}})\hat{\mathbf{D}}, \qquad \hat{\phi} = \frac{1}{n-p}\sum_i (Y_i - \hat{\mu}_i)^2 / V_i(\hat{\mu}_i).$$

Since $\hat{\boldsymbol{\beta}}$ is a consistent estimator of $\boldsymbol{\beta}$, thus, by Slutsky's theorem we will have $D_n = \hat{\phi}^{-1}\xi_n'\mathbf{F}(\mathbf{F}'\hat{\Sigma}^{-1}\mathbf{F})^{-1}\mathbf{F}'\xi_n$
$\xrightarrow[n\to\infty]{D} \chi_q^2(\Delta)$.

Proof. See Proofs. □

Based on the above conditions and under the local alternative K_n the following theorems hold.

Theorem 3.1. *If (1) and (2) hold, then under K_n, as $n \to \infty$ the ADB of the estimators are respectively*

$$\begin{aligned} ADB(\hat{\boldsymbol{\beta}}) &= \mathbf{0}, \\ ADB(\tilde{\boldsymbol{\beta}}) &= -\boldsymbol{\delta}, \\ ADB(\hat{\boldsymbol{\beta}}^{PT}) &= -\boldsymbol{\delta} H_{q+2}(\chi_q^2(\alpha); \Delta), \end{aligned}$$

with $\Delta = \phi^{-1}\boldsymbol{\omega}'(\mathbf{F}'\boldsymbol{\Sigma}^{-1}\mathbf{F})^{-1}\boldsymbol{\omega}$, $H_\nu(x;\Delta)$ is the distribution function of a non-central chi-square with ν degrees of freedom and non-centrality parameter Δ, and

$$E\big(\chi_\nu^{-2j}(\Delta)\big) = \int_0^\infty \chi^{-2j} dH_\nu(x;\Delta).$$

Proof. See Proofs. □

Since the bias expressions are not in scalar form, we convert them to quadratic form. Therefore we define the asymptotic quadratic distributional bias (AQDB) of an estimator as follows

$$AQDB(\boldsymbol{\beta}^*) = (ADB(\boldsymbol{\beta}^*))'\phi^{-1}\boldsymbol{\Sigma}(ADB(\boldsymbol{\beta}^*)).$$

Theorem 3.2. *Suppose that the assumptions of Theorem 3.1 hold, then under K_n, as $n \to \infty$ the AQDB of the estimators are*

$$\begin{aligned} AQDB(\hat{\boldsymbol{\beta}}) &= \mathbf{0}, \\ AQDB(\tilde{\boldsymbol{\beta}}) &= \phi^{-1}\boldsymbol{\delta}'\Sigma\boldsymbol{\delta} = \Delta, \\ AQDB(\hat{\boldsymbol{\beta}}^{PT}) &= \Delta\big(H_{q+2}(\chi_q^2(\alpha);\Delta)\big)^2. \end{aligned}$$

Proof. See Proofs. □

The asymptotic dispersion matrices of the estimators are given below.

Theorem 3.3. *Suppose that the assumptions of Theorem 3.1 hold, then under K_n, as $n \to \infty$ the asymptotic covariance matrices of the estimators are*

$$\begin{aligned} \boldsymbol{\Gamma}(\hat{\boldsymbol{\beta}}) &= \phi\boldsymbol{\Sigma}^{-1}, \\ \boldsymbol{\Gamma}(\tilde{\boldsymbol{\beta}}) &= \phi\boldsymbol{\Sigma}^* + \boldsymbol{\delta}\boldsymbol{\delta}', \\ \boldsymbol{\Gamma}(\hat{\boldsymbol{\beta}}^{PT}) &= \phi\boldsymbol{\Sigma}^{-1} - \phi(\boldsymbol{\Sigma}^{-1} - \boldsymbol{\Sigma}^*)H_{q+2}(\chi_q^2(\alpha);\Delta) \\ &\quad + \boldsymbol{\delta}\boldsymbol{\delta}'\Big\{2H_{q+2}(\chi_q^2(\alpha);\Delta) - H_{q+4}(\chi_q^2(\alpha);\Delta)\Big\}. \end{aligned}$$

Proof. See Proofs. □

Theorem 3.4. *Suppose that the assumptions of theorem 4.1 hold, then under K_n, the ADR of the estimators are*

$$\begin{aligned} R(\hat{\boldsymbol{\beta}};\mathbf{M}) &= \phi\ tr(\mathbf{M}\boldsymbol{\Sigma}^{-1}), \\ R(\tilde{\boldsymbol{\beta}};\mathbf{M}) &= \phi\ tr(\mathbf{M}\boldsymbol{\Sigma}^{-1}) - \phi\ tr(\mathbf{A}_{11}) + \boldsymbol{\delta}'\mathbf{M}\boldsymbol{\delta}, \\ R(\hat{\boldsymbol{\beta}}^{PT};\mathbf{M}) &= \phi\ tr(\mathbf{M}\boldsymbol{\Sigma}^{-1}) - \phi\ tr(\mathbf{A}_{11})H_{q+2}(\chi_q^2(\alpha);\Delta) \\ &\quad + \boldsymbol{\delta}'\mathbf{M}\boldsymbol{\delta}\Big\{2H_{q+2}(\chi_q^2(\alpha);\Delta) - H_{q+4}(\chi_q^2(\alpha);\Delta)\Big\}, \end{aligned}$$

with $\mathbf{A}_{11} = \mathbf{M}\boldsymbol{\Sigma}^{-1}\mathbf{F}(\mathbf{F}'\boldsymbol{\Sigma}^{-1}\mathbf{F})^{-1}\mathbf{F}'\boldsymbol{\Sigma}^{-1}$.

Proof. See Proofs. □

3.1. *Bias and Risk Comparison*

The quadratic biases of $\tilde{\boldsymbol{\beta}}$ and $\hat{\boldsymbol{\beta}}^{PT}$ are functions of Δ. Under H_0, i.e., when $\Delta = 0$, $\tilde{\boldsymbol{\beta}}$ and $\hat{\boldsymbol{\beta}}^{PT}$ are unbiased estimators, however as Δ moves away from 0, the $AQDB(\tilde{\boldsymbol{\beta}})$ becomes unbounded function of Δ and $AQDB(\hat{\boldsymbol{\beta}}^{PT})$ will be less than $AQDB(\tilde{\boldsymbol{\beta}})$ for all values of Δ since $H_{q+2}(\chi_q^2(\alpha);\Delta)$ lies between 0 and 1.

By comparing the risks of the estimators, we see that, as Δ moves away from 0, the $ADR(\tilde{\boldsymbol{\beta}})$ becomes unbounded. Also, by comparing the risk of $\hat{\boldsymbol{\beta}}^{PT}$ and $\tilde{\boldsymbol{\beta}}$, it can be easily shown that, beyond the small values of Δ ($\Delta \in [0, c]$), $ADR(\hat{\boldsymbol{\beta}}^{PT}) < ADR(\tilde{\boldsymbol{\beta}})$. As Δ increases the $ADR(\hat{\boldsymbol{\beta}}^{PT})$ will increase and reaches the $ADR(\hat{\boldsymbol{\beta}})$ from below. More importantly, it is a bounded function of Δ. However, under the null hypothesis:

$$ADR(\tilde{\boldsymbol{\beta}}) < ADR(\hat{\boldsymbol{\beta}}^{PT}) < ADR(\hat{\boldsymbol{\beta}}).$$

4. Monte Carlo Simulation

In this section, we provide a Monte Carlo simulation study to investigate the performance of the proposed estimators with different numbers of explanatory variables. In this study, we considered the following model:

$$log(\lambda_i) = \mathbf{x}_i'\boldsymbol{\beta}, \quad i = 1, 2, \ldots, n,$$

where $\mathbf{x}_i' = (1, x_{1i}, x_{2i}, \ldots, x_{pi})$, $\lambda_i = E(y_i|\mathbf{x}_i)$, and $y_i's$ are observations from an over-dispersed poisson model. In order to generate over-dispersed poisson observations with mean λ_i and variance $\phi\lambda_i$, we considered a $NB(r_i, p)$ with

$$r_i = \frac{\lambda_i}{\phi - 1} \text{ and } p = \frac{1}{\phi}, \quad i = 1, 2, \ldots, n,$$

where $\lambda_i = e^{\mathbf{x}_i'\boldsymbol{\beta}}$ and $x_{si} = t_s^2 + \nu_i$ with t_s and ν_i being i.i.d $N(0,1)$ for all $s = 1,\ldots,p$ and $i = 1,\ldots,n$. Also, in the simulation we considered $\phi = 2$. Our sampling experiment consists of various combinations of sample sizes, i.e., $n = 50, 100$ and 150. For each n, we generate 5000 samples using the above model. We also use the 10-fold cross validation method to estimate the tuning parameter λ to compute lasso. Furthermore, we use the aod-package[14] in R statistical software to fit the above log linear model to account for the over-dispersed poisson model. In our simulation, we consider the UPI in the following format:
$\mathbf{F}' = (\mathbf{0}, \mathbf{I})$ where $\mathbf{I}_{p_2\times p_2}$ is the identity matrix, and $\mathbf{0}_{p_2\times p_1}$ is the matrix of 0s and $\mathbf{d}_{p_2\times 1} = \mathbf{0}$. Also, we set the regression coefficients of $\boldsymbol{\beta} = (\boldsymbol{\beta}_1', \boldsymbol{\beta}_2')'$ to $\boldsymbol{\beta} = (\boldsymbol{\beta}_1', \mathbf{0}')$ with $\beta_j = 0$, for $j = p_1+1,\ldots,p$ with $p = p_1 + p_2$ for the following case:
$\boldsymbol{\beta}_1 = (1,1,1)$ and $\boldsymbol{\beta}_2 = \mathbf{0}_{p_2\times_1}$ with dimensions $p_2 = 3, 5, 7$.
Now we define the parameter $\Delta = \|\boldsymbol{\beta} - \boldsymbol{\beta}^*\|^2$, where $\boldsymbol{\beta}^* = (\boldsymbol{\beta}_1', \mathbf{0}')'$ and $\|\ .\ \|$ is the Euclidian norm. The objective is to investigate the behavior of the estimators for $\Delta \geq 0$. In order to do this, further samples are generated from those distributions (i.e. for different Δ between 0 and 2). To produce different values of Δ, different values of $\boldsymbol{\beta}_2$ are chosen. The criterion for comparing the performance of the estimators of $\boldsymbol{\beta}_1$ is based on the mean squared error (MSE). The relative MSE of the estimators $\tilde{\boldsymbol{\beta}}, \hat{\boldsymbol{\beta}}^{PT}$, and $\hat{\boldsymbol{\beta}}_{lasso}$ have been numerically calculated with respect to $\hat{\boldsymbol{\beta}}$ using the R statistical software. The relative mean squared error (RMSE) of the other estimators to the unrestricted estimator $\hat{\boldsymbol{\beta}}$ is defined by

$$RMSE(\hat{\boldsymbol{\beta}} : \hat{\boldsymbol{\beta}}^{\star}) = \frac{MSE(\hat{\boldsymbol{\beta}})}{MSE(\hat{\boldsymbol{\beta}}^{\star})},$$

where $\hat{\boldsymbol{\beta}}^{\star}$ can be any of $\tilde{\boldsymbol{\beta}}, \hat{\boldsymbol{\beta}}^{PT}$, and $\hat{\boldsymbol{\beta}}_{lasso}$. It is obvious that a RMSE larger than one indicates the degree of superiority of the estimator $\hat{\boldsymbol{\beta}}^{\star}$ over $\hat{\boldsymbol{\beta}}$.

The performance of lasso and the UE are independent of the parameter Δ. These estimators do not take advantage of the fact that the regression parameter lies in a subspace and may be at a disadvantage when $\Delta > 0$. Figures 1-3 portray the relative performance of the suggested estimators. We summarize our findings as follows:

(i) both lasso and PTE outperform the UE in all simulation cases. When UPI is correctly specified the PTE outperforms lasso, indicating that PTE has a lower MSE compared to the lasso

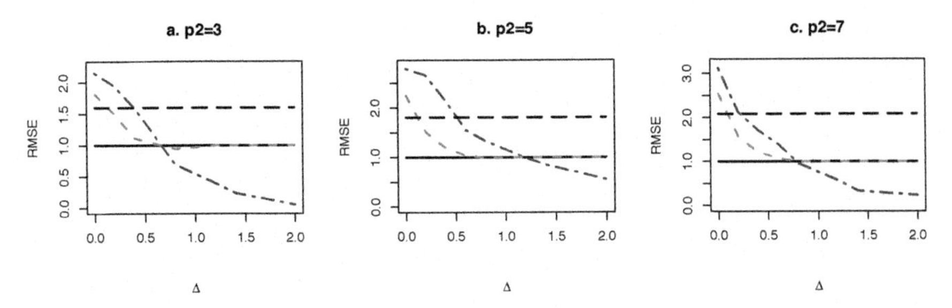

Fig. 1. Relative MSE of the estimators for various p_2 when $n = 50$. "- - -" denotes the PTE, "– · – ·" denotes the RE, "—" denotes the UE, and "– – –" denotes the lasso estimator.

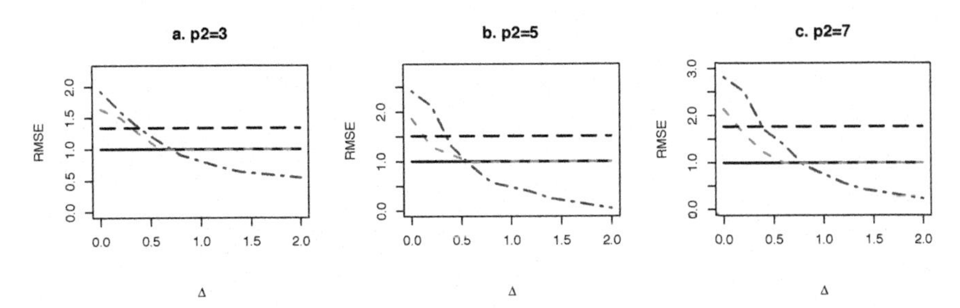

Fig. 2. Relative MSE of the estimators for various p_2 when $n = 100$. "- - -" denotes the PTE, "– · – ·" denotes the RE, "—" denotes the UE, and "– – –" denotes the lasso estimator.

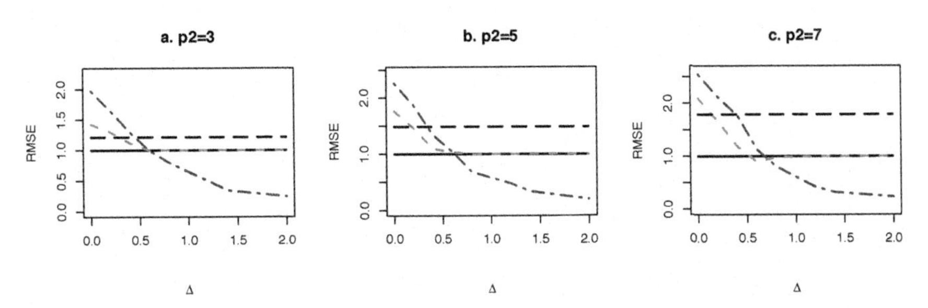

Fig. 3. Relative MSE of the estimators for various p_2 when $n = 150$. "- - -" denotes the PTE, "– · – ·" denotes the RE, "—" denotes the UE, and "– – –" denotes the lasso estimator.

(ii) for smaller values of Δ, the RE performs better than the PTE and lasso

(iii) however, beyond the small interval near the null hypothesis the PTE outperforms the RE and has lower MSE than RE

(iv) the RE performs best only when Δ is small. For large values of Δ, it becomes inconsistent and its RMSE converges to 0. Again, if a submodel is nearly correctly specified, then the RE is the optimal one.

5. Conclusions

In this study, we suggested an estimation strategy for estimating regression parameters using QL strategy. In addition, we suggested pretest and absolute penalty estimators. The simulation and analytical results demonstrate that the RE dominates the other estimators under a correctly specified model. However, the improvement will be diminishing as restriction becomes incorrect. On the other hand, the PTE performs better than the UE at and near the restriction and its risk function is bounded in Δ.

Numerical results demonstrated that the lasso estimator is better than the UE for all values of p_2. Also the results showed that the PTE has lower MSE than lasso and outperforms lasso in all the cases. But when the restriction is true, the RE is superior to all the other estimation rules. Finally, it would be interesting to investigate the relative performance of improved pretest estimators with some other APE estimators. We leave this for further investigation.

The proposed estimation strategy can be extended in various directions to more complex problems. Research on the statistical implications of proposed and related estimators is ongoing. Pretest, APE and likelihood-based methods continue to be extremely useful tools for efficient estimation.

Acknowledgments

This research was supported by the Natural Sciences and the Engineering Council of Canada (NSERC) and the Canadian Institute of Health Sciences(CIHR). We thank two referees and editor Tõnu Kollo for their constructive comments and suggestions, which led to significant improvement in the presentation of this paper.

6. Proofs

Proof of Corollary 3.1. By using Proposition 3.1, under local alternative, we have $\xi_n \xrightarrow[n\to\infty]{D} \mathcal{N}_p(\boldsymbol{\delta}, \phi\ \mathbf{C})$ where $\mathbf{C} = \phi(\boldsymbol{\Sigma}^{-1} - \boldsymbol{\Sigma}^*) = \phi\boldsymbol{\Sigma}^{-1}\mathbf{F}(\mathbf{F}'\boldsymbol{\Sigma}^{-1}\mathbf{F})^{-1}\mathbf{F}'\boldsymbol{\Sigma}^{-1}$. Consider $\mathbf{A} = \phi^{-1}\mathbf{F}(\mathbf{F}'\boldsymbol{\Sigma}^{-1}\mathbf{F})^{-1}\mathbf{F}'$ which is a symmetric matrix, one can verify that the following conditions hold:

1) $(\mathbf{AC})^2 = \mathbf{AC}$, 2) $\boldsymbol{\delta}'(\mathbf{AC})^2 = \boldsymbol{\delta}'\mathbf{AC}$, 3) $\boldsymbol{\delta}'\,\mathbf{ACA}\boldsymbol{\delta} = \boldsymbol{\delta}'\,\mathbf{A}\boldsymbol{\delta}$, 4) $r(\mathbf{AC}) = q$
under our regularity conditions and using Theorem 4 in Styan[15], we get $\phi^{-1}\xi_n'\mathbf{F}(\mathbf{F}'\boldsymbol{\Sigma}^{-1}\mathbf{F})^{-1}\mathbf{F}'\xi_n \sim \chi_q^2(\Delta)$ where $\Delta = \phi^{-1}\boldsymbol{\delta}'\mathbf{A}\boldsymbol{\delta} = \phi^{-1}\boldsymbol{\omega}'(\mathbf{F}'\boldsymbol{\Sigma}^{-1}\mathbf{F})^{-1}\boldsymbol{\omega}$. □

Proof of Proposition 3.1. Since ζ_n and ξ_n are linear functions of $\hat{\boldsymbol{\beta}}$, they are also asymptotically normally distributed.

$$
\begin{aligned}
E(\zeta_n) &= E \lim_{n\to\infty} \sqrt{n}(\tilde{\boldsymbol{\beta}} - \boldsymbol{\beta}) \\
&= E \lim_{n\to\infty} \sqrt{n}(\hat{\boldsymbol{\beta}} - \boldsymbol{\Sigma}^{-1}\mathbf{F}(\mathbf{F}\boldsymbol{\Sigma}^{-1}\mathbf{F})^{-1}(\mathbf{F}'\hat{\boldsymbol{\beta}} - \mathbf{d}) - \boldsymbol{\beta}) \\
&= E \lim_{n\to\infty} \sqrt{n}(\hat{\boldsymbol{\beta}} - \boldsymbol{\beta} - \boldsymbol{\Sigma}^{-1}\mathbf{F}(\mathbf{F}\boldsymbol{\Sigma}^{-1}\mathbf{F})^{-1}(\mathbf{F}'\hat{\boldsymbol{\beta}} - \mathbf{F}'\boldsymbol{\beta} + \mathbf{F}'\boldsymbol{\beta} - \mathbf{d}) \\
&= E \lim_{n\to\infty} \sqrt{n}(\hat{\boldsymbol{\beta}} - \boldsymbol{\beta}) \\
&- E \lim_{n\to\infty} \sqrt{n}\{\boldsymbol{\Sigma}^{-1}\mathbf{F}(\mathbf{F}'\boldsymbol{\Sigma}^{-1}\mathbf{F})^{-1}[\mathbf{F}'(\hat{\boldsymbol{\beta}} - \boldsymbol{\beta}) + \frac{\boldsymbol{\omega}}{\sqrt{n}}]\} \\
&= E(\lim_{n\to\infty} \varrho_n) - \boldsymbol{\Sigma}^{-1}\mathbf{F}(\mathbf{F}'\boldsymbol{\Sigma}^{-1}\mathbf{F})^{-1}(E \lim_{n\to\infty}(\mathbf{F}'\varrho_n) + \boldsymbol{\omega}) \\
&= -\boldsymbol{\Sigma}^{-1}\mathbf{F}(\mathbf{F}'\boldsymbol{\Sigma}^{-1}\mathbf{F})^{-1}\boldsymbol{\omega} = -\boldsymbol{\delta} \\
V(\zeta_n) &= Var(\sqrt{n}(\tilde{\boldsymbol{\beta}} - \boldsymbol{\beta})) = Var(\varrho_n - \boldsymbol{\Sigma}^{-1}\mathbf{F}(\mathbf{F}'\boldsymbol{\Sigma}^{-1}\mathbf{F})^{-1}\mathbf{F}'\varrho_n) \\
&= \boldsymbol{\Sigma}^{-1} + \boldsymbol{\Sigma}^{-1}\mathbf{F}(\mathbf{F}'\boldsymbol{\Sigma}^{-1}\mathbf{F})^{-1}\mathbf{F}'\boldsymbol{\Sigma}^{-1}\mathbf{F}(\mathbf{F}'\boldsymbol{\Sigma}^{-1}\mathbf{F})^{-1}\mathbf{F}'\boldsymbol{\Sigma}^{-1} \\
&- 2\boldsymbol{\Sigma}^{-1}\mathbf{F}(\mathbf{F}'\boldsymbol{\Sigma}^{-1}\mathbf{F})^{-1}\mathbf{F}'\boldsymbol{\Sigma}^{-1} = \boldsymbol{\Sigma}^{-1} - \boldsymbol{\Sigma}^{-1}\mathbf{F}(\mathbf{F}'\boldsymbol{\Sigma}^{-1}\mathbf{F})^{-1}\mathbf{F}'\boldsymbol{\Sigma}^{-1}.
\end{aligned}
$$

In a similar way, one can achieve the asymptotic results of ξ_n. Now the joint distribution of (ϱ_n, ζ_n) and (ζ_n, ξ_n) will be asymptotically normal as well. □

We present the following Lemma below, which will enable us to derive the results of Theorems 3.1-3.4 in this paper.

Lemma 1. Let $\mathbf{x}$ be a q-dimensional normal vector distributed as $N_q(\boldsymbol{\mu}_{\mathbf{x}}, \boldsymbol{\Sigma}_q)$, then, for a measurable function of φ, we have

$$
\begin{aligned}
E[\mathbf{x}\varphi(\mathbf{x}'\mathbf{x})] &= \boldsymbol{\mu}_{\mathbf{x}} E[\varphi(\chi^2_{q+2}(\Delta))] \\
E[\mathbf{x}\mathbf{x}'\varphi(\mathbf{x}'\mathbf{x})] &= \boldsymbol{\Sigma}_q E[\varphi(\chi^2_{q+2}(\Delta))] + \boldsymbol{\mu}_{\mathbf{x}}\boldsymbol{\mu}_{\mathbf{x}}' E[\varphi(\chi^2_{q+4}(\Delta))],
\end{aligned}
$$

where $\Delta = \boldsymbol{\mu}_{\mathbf{x}}'\boldsymbol{\Sigma}_q^{-1}\boldsymbol{\mu}_{\mathbf{x}}$. The proof can be found in Judge and Bock.[16]

Now, we provide the proof for the Theorems. Proposition 3.1 and Lemma 1 are used to prove the results of the Theorems.

Proof of Theorem 3.1. Here, we provide the proof of bias expressions. From Proposition 3.1, we get directly the statements $ADB(\hat{\boldsymbol{\beta}}) = \mathbf{0}$, and $ADB(\hat{\boldsymbol{\beta}}) = -\boldsymbol{\delta}$. The ADB of the pretest estimator is as follows:

$$\begin{aligned}
ADB(\hat{\boldsymbol{\beta}}^{PT}) &= E \lim_{n\to\infty} \sqrt{n}(\hat{\boldsymbol{\beta}}^{PT} - \boldsymbol{\beta}) \\
&= E \lim_{n\to\infty} \sqrt{n}[\hat{\boldsymbol{\beta}} - (\hat{\boldsymbol{\beta}} - \tilde{\boldsymbol{\beta}})I(D_n < \chi_q^2(\alpha)) - \boldsymbol{\beta}] \\
&= E \lim_{n\to\infty} [\varrho_n - \xi_n\ I(D_n < \chi_q^2(\alpha))] \\
&= -E \lim_{n\to\infty} [\xi_n\ I(D_n < \chi_q^2(\alpha))] \\
&= -\boldsymbol{\delta} E \lim_{n\to\infty} [I(\chi_{q+2}^2(\Delta) < \chi_q^2(\alpha))] = -\boldsymbol{\delta} H_{q+2}(\chi_q^2(\alpha); \Delta). \quad \square
\end{aligned}$$

Proof of Theorem 3.2. One can easily derive the results for Theorem 3.2 by following the definition of AQDB. □

Proof of Theorem 3.3. The asymptotic covariance matrices of the estimators are:

$$\begin{aligned}
\boldsymbol{\Gamma}(\hat{\boldsymbol{\beta}}) &= E \lim_{n\to\infty} (n(\hat{\boldsymbol{\beta}} - \boldsymbol{\beta})(\hat{\boldsymbol{\beta}} - \boldsymbol{\beta})') = E \lim_{n\to\infty} (\varrho_n \varrho_n') \\
&= Var(\varrho_n) + E(\varrho_n)E(\varrho_n)' \\
&= \phi \boldsymbol{\Sigma}^{-1} \\
\boldsymbol{\Gamma}(\tilde{\boldsymbol{\beta}}) &= E \lim_{n\to\infty} (n(\tilde{\boldsymbol{\beta}} - \boldsymbol{\beta})(\tilde{\boldsymbol{\beta}} - \boldsymbol{\beta})') = E \lim_{n\to\infty} (\zeta_n \zeta_n') \\
&= Var(\zeta_n) + E(\zeta_n)E(\zeta_n)' \\
&= \phi \boldsymbol{\Sigma}^* + \boldsymbol{\delta}\boldsymbol{\delta}' \\
\boldsymbol{\Gamma}(\hat{\boldsymbol{\beta}}^{PT}) &= E \lim_{n\to\infty} (n(\hat{\boldsymbol{\beta}}^{PT} - \boldsymbol{\beta})(\hat{\boldsymbol{\beta}}^{PT} - \boldsymbol{\beta})') \\
&= E \lim_{n\to\infty} \sqrt{n}\Big(\hat{\boldsymbol{\beta}} - (\hat{\boldsymbol{\beta}} - \tilde{\boldsymbol{\beta}})I(D_n < \chi_q^2(\alpha)) - \boldsymbol{\beta}\Big) \\
&\times \sqrt{n}\Big(\hat{\boldsymbol{\beta}} - (\hat{\boldsymbol{\beta}} - \tilde{\boldsymbol{\beta}})I(D_n < \chi_q^2(\alpha)) - \boldsymbol{\beta}\Big)' \\
&= E \lim_{n\to\infty} \Big(\varrho_n - \xi_n\ I(D_n < \chi_q^2(\alpha))\Big)\Big(\varrho_n - \xi_n\ I(D_n < \chi_q^2(\alpha))\Big)' \\
&= E \lim_{n\to\infty} \Big(\rho_n \rho_n' - \varrho_n \xi_n'\ I(D_n < \chi_q^2(\alpha)) - \xi_n \rho_n'\ I(D_n < \chi_q^2(\alpha)) \\
&+ \xi_n \xi_n'\ I(D_n < \chi_q^2(\alpha))\Big).
\end{aligned}$$

Note that, by using Lemma 1 and the formula for a conditional mean of a bivariate normal, we have

$$
\begin{aligned}
&E \lim_{n\to\infty} [\xi_n \rho_n' \ I(D_n < \chi_q^2(\alpha))] = E \lim_{n\to\infty} (E(\xi_n \rho_n' \ I(D_n < \chi_q^2(\alpha))|\xi_n)) \\
&= E \lim_{n\to\infty} \Big(\xi_n [E(\rho_n) + \phi(\boldsymbol{\Sigma}^{-1} - \boldsymbol{\Sigma}^*)\phi^{-1}(\boldsymbol{\Sigma}^{-1} - \boldsymbol{\Sigma}^*)^{-1} \\
&\quad (\xi_n - E(\xi_n))]' I(D_n < \chi_q^2(\alpha))\Big) \\
&= E \lim_{n\to\infty} \Big(\xi_n [\xi_n' - \boldsymbol{\delta}'] I(D_n < \chi_q^2(\alpha))\Big) \\
&= E \lim_{n\to\infty} (\xi_n \xi_n' \ I(D_n < \chi_q^2(\alpha))) - E \lim_{n\to\infty} (\xi_n \boldsymbol{\delta}' \ I(D_n < \chi_q^2(\alpha))) \\
&= \phi(\boldsymbol{\Sigma}^{-1} - \boldsymbol{\Sigma}^*) E(I(\chi_{q+2}^2(\Delta) < \chi_q^2(\alpha))) + \boldsymbol{\delta}\boldsymbol{\delta}' E(I(\chi_{q+4}^2(\Delta) < \chi_q^2(\alpha))) \\
&-\boldsymbol{\delta}\boldsymbol{\delta}' E(I(\chi_{q+2}^2(\Delta) < \chi_q^2(\alpha))) \\
&= \phi(\boldsymbol{\Sigma}^{-1} - \boldsymbol{\Sigma}^*) H_{q+2}(\chi_q^2(\alpha); \Delta) + \boldsymbol{\delta}\boldsymbol{\delta}' H_{q+4}(\chi_q^2(\alpha); \Delta) \\
&-\boldsymbol{\delta}\boldsymbol{\delta}' H_{q+2}(\chi_q^2(\alpha); \Delta).
\end{aligned}
$$

Therefore,

$$
\begin{aligned}
\boldsymbol{\Gamma}(\hat{\boldsymbol{\beta}}^{PT}) &= \phi\boldsymbol{\Sigma}^{-1} - 2\phi(\boldsymbol{\Sigma}^{-1} - \boldsymbol{\Sigma}^*) H_{q+2}(\chi_q^2(\alpha); \Delta) - 2\boldsymbol{\delta}\boldsymbol{\delta}' H_{q+4}(\chi_q^2(\alpha); \Delta) \\
&\quad + 2\boldsymbol{\delta}\boldsymbol{\delta}' H_{q+2}(\chi_q^2(\alpha); \Delta) + \phi(\boldsymbol{\Sigma}^{-1} - \boldsymbol{\Sigma}^*) H_{q+2}(\chi_q^2(\alpha); \Delta) \\
&\quad + \boldsymbol{\delta}\boldsymbol{\delta}' H_{q+4}(\chi_q^2(\alpha); \Delta) \\
&= \phi\boldsymbol{\Sigma}^{-1} - \phi(\boldsymbol{\Sigma}^{-1} - \boldsymbol{\Sigma}^*) H_{q+2}(\chi_q^2(\alpha); \Delta) - \boldsymbol{\delta}\boldsymbol{\delta}' H_{q+4}(\chi_q^2(\alpha); \Delta) \\
&\quad + 2\boldsymbol{\delta}\boldsymbol{\delta}' H_{q+2}(\chi_q^2(\alpha); \Delta)
\end{aligned}
$$

□

Proof of Theorem 3.4. As previously defined the ADR of an estimator is $R(\boldsymbol{\beta}^*, \mathbf{M}) = tr(\mathbf{M}\boldsymbol{\Gamma})$, where $\boldsymbol{\Gamma}$ is the asymptotic covariance of the estimators.

$$
\begin{aligned}
R(\hat{\boldsymbol{\beta}}; \mathbf{M}) &= \mathrm{tr}\{\mathbf{M}\phi\boldsymbol{\Sigma}^{-1}\} = \phi \ \mathrm{tr}(\mathbf{M}\boldsymbol{\Sigma}^{-1}), \\
R(\tilde{\boldsymbol{\beta}}; \mathbf{M}) &= \mathrm{tr}\{\phi\mathbf{M}\boldsymbol{\Sigma}^* + \mathbf{M}\boldsymbol{\delta}\boldsymbol{\delta}'\} \\
&= \phi \ \mathrm{tr}(\mathbf{M}\boldsymbol{\Sigma}^{-1}) - \phi \ \mathrm{tr}(\mathbf{A}_{11}) + \boldsymbol{\delta}'\mathbf{M}\boldsymbol{\delta}, \\
R(\hat{\boldsymbol{\beta}}^{PT}; \mathbf{M}) &= \mathrm{tr}\Big\{\phi\mathbf{M}\boldsymbol{\Sigma}^{-1} - \phi\mathbf{M}(\boldsymbol{\Sigma}^{-1} - \boldsymbol{\Sigma}^*) H_{q+2}(\chi_q^2(\alpha); \Delta) \\
&\quad + \mathbf{M}\boldsymbol{\delta}\boldsymbol{\delta}'(2H_{q+2}(\chi_q^2(\alpha); \Delta) - H_{q+4}(\chi_q^2(\alpha); \Delta))\Big\} \\
&= \phi \ \mathrm{tr}(\mathbf{M}\boldsymbol{\Sigma}^{-1}) - \phi \ \mathrm{tr}(\mathbf{A}_{11}) H_{q+2}(\chi_q^2(\alpha); \Delta) \\
&\quad + \boldsymbol{\delta}'\mathbf{M}\boldsymbol{\delta}\Big\{2H_{q+2}(\chi_q^2(\alpha); \Delta) - H_{q+4}(\chi_q^2(\alpha); \Delta)\Big\},
\end{aligned}
$$

with $\mathbf{A}_{11} = \mathbf{M}\boldsymbol{\Sigma}^{-1}\mathbf{F}(\mathbf{F}'\boldsymbol{\Sigma}^{-1}\mathbf{F})^{-1}\mathbf{F}'\boldsymbol{\Sigma}^{-1}$. □

References

1. R. W. M. Wedderburn, *Biometrika* **61** (1974).
2. P. McCullagh, *Annals of Statistics* **11** (1983).
3. C. C. Heyde, *Quasi-Likelihood And Its Application: A General Approach to Optimal Parameter Estimation* (Springer-Verlag, New York, 1997).
4. S. E. Ahmed, *Lecture Notes in Statistics* **148**, 103 (2001), Editors: Ahmed, S. E. and Reid, N. Springer-Verlag: New York.
5. S. E. Ahmed, A. Hussein and S. Nkurunziza, *Statistics & Probability Letters* **80**, 726 (2010).
6. S. E. Ahmed and S. Liu, *Linear Algebra and its Applications* **430**, 2734 (2009).
7. B. Efron, T. Hastie, I. Johnstone and R. Tibshirani, *Annals of Statistics* **33**, 407 (2004).
8. M. Y. Park and T. Hastie, *Journal of Royal Statistical Society B* **69** (2007).
9. S. E. Ahmed, K. A. Doksum, S. Hossein and J. You, *Australian & New Zealand Journal of Statistics* **49**, 435 (2007).
10. S. Fallahpour, S. E. Ahmed and K. A. Doksum, *Journal of Applied Stochastic Models in Business and Industry* **28**, 236 (2012).
11. J. Friedman, T. Hastie and R. Tibshirani, *Journal of Statistical Software* **32**, 1 (2010).
12. J. Friedman, T. Hastie and R. Tibshirani (2009), R package version 1.1-4, `http://CRAN.R-project.org/package=glmnet`.
13. R Development Core Team, R: A language and environment for statistical computing. R Foundation for Statistical Computing, vienna, austria (2011), `http://www.R-project.org/`.
14. M. Lesnoff and R. Lancelot (2012), R package version 1.1-4, `http://cran.r-project.org/package=aod`.
15. G. P. H. Styan, *Biometrika* **57** (1970).
16. G. G. Judge and M. E. Bock, *The Statistical Implication of Pre-test and Stein-rule Estimators in Econometrics*, 1978).

MAXIMUM LIKELIHOOD ESTIMATES FOR MARKOV-ADDITIVE PROCESSES OF ARRIVALS BY AGGREGATED DATA

A. M. ANDRONOV

Department of Mathematical Methods and Modelling,
Transport and Telecommunication Institute,
Riga, LV-1019, Latvia
E-mail: lora@.mailbox.riga.lv

A problem of parameter estimation for Markov-Additive Processes of arrivals is considered. It is supposed that empirical data are aggregated: only overall numbers of arrivals of various classes are observed in a long time interval. Maximum likelihood method is used for estimation: score function is derived and the gradient method is used for the optimization. Numerical example is considered.

Keywords: Gradient method; Kronecker product; Markov-Additive process; score function.

1. Introduction

Lately great attention has been payd to a dependence between interarrival times of various flows. Usually the one is described by the so-called Batch Markovian Arrival Process and its modifications (see Refs. 1-8) or by Markov-Additive Processes of arrivals (see Ref. 9). More generally, Markov-modulated processes are under consideration. They are widely used in environmental, medical, industrial, and sociological studies.

Statistical inference for Markov-modulated processes have received little attention. More often parameter estimation for Markov-modulated Poisson process (MMPP) is considered. Davison and Ramesh[10] applied numerical optimizing for a discretized version of the process. Turin[11] proposed an Expectation Maximization algorithm for finding maximum likelihood estimates. Scott[7] provided a Bayesian method for inferring the parameters of a stationary two-state MMPP. Scott and Smyth[8] extend Scott[7] to the non-stationary case with an arbitrary number of states.

Further we formulate the considered problem in terms of Markov-Additive Processes (MAP) of arrivals that is more general. MAP of arrivals supposes that various classes of arrived calls take place. We have independent observations about numbers of various arrivals for a long time-period; it is the so-called aggregated data case. Their distribution is approximated by a multi-variate normal distribution. Our aim is to calculate maximum likelihood estimates for parameters of the considered MAP of arrivals.
The paper is organized as follows. In Section 2 Markov-Additive Process of arrivals is described. A problem of parameter estimation is formulated in Sec. 3. Sections 4 and 5 are devoted to derivative calculation for the expectation and the covariance matrix correspondingly. Score function is presented in Sec. 6. Problem of parameter identification is discussed in Sec. 7. Numerical example in Sec. 8 illustrates main results. Section 9 concludes the paper.

2. Markov-Additive Process of arrivals

A definition of Markov-Additive Process of arrivals has been given by Pacheco *et al*[9] (see Chapter 6). We consider the following simplification of this process.
Let $N = \{0, 1, ...\}, r$ be a positive integer, and E be a countable set. A considered process $(X, J) = \{(X(t), J(t)), t \geq 0\}$ on the state space $N^r \times E$ satisfies the following two conditions:

(a) (X, J) is *a time-homogeneous Markov chain*;
(b) For $s, t \geq 0$, *the conditional distribution of* $(X(s + t) - X(s), J(s + t))$ *given* $J(s)$ *depends only on* t.

Within the current framework, the increments of X are associated to arrival calls. Different (namely r) classes of arrivals are possible, so $X_i(t)$ equals total call number of class i arrivals in $(0, t]$, $i = 1, 2, ..., r$. One arrival can contain various numbers of calls of each class. We call X the *arrival component* of (X, J), and J - the *Markov component* of (X, J). Let $m = |E|$. Since (X, J) is time-homogeneous, it sufficient to give, for $j, k \in E$ and $n, n' \in N^r$, the transition rate from (n', j) to $(n' + n, k)$, which we denote by $\lambda_{j,k}(n)$.
Whenever the Markov component J is in the state j, the following two types of transitions in (X, J) may occur:

(a) *arrivals without change of state* in $j \in E$ occur at rate $\lambda_{j,j}(n), n > 0$;

(b) *changes of state in J without arrivals occur at rate $\lambda_{j,k}(0), k \in E, j \neq k$. Arrivals with a change of state in J are impossible.*

We denote $\Lambda_n = (\lambda_{j,k}(n)),\ n \in N^r$, so for $n > 0$ matrix Λ_n is a diagonal one: $\Lambda_n = diag(\lambda_{j,j}(n) : j = 1, ..., m)$. (X, J) is *a simple* MAP of arrivals if $\Lambda_n = 0$ when $n_i > 1$ for some i-th component of n. (X, J) is *a univariate* MAP of arrivals if $r = 1$. In this case we speak about a Markov-Modulated Poisson Process (MMPP).
Now we introduce some additional notation. *Let γ_j be total rate of change in J for state $j \in E$:*

$$\gamma_j = \sum_{\forall k} \lambda_{j,k}(0).$$

Further, let

$$\Gamma = diag(\gamma_j), \Lambda = \sum_{n>0} \Lambda_n, Q = (q_{j,k}) = \Lambda_0 - \Gamma, \overline{\sum_p}^{i} = \sum_{n>0} (n_i)^p \Lambda_n.$$

Let us comment the introduced notation:

(a) Λ is the matrix of transition rates in J associated with arrivals; it is a diagonal matrix;
(b) Λ_0 is the matrix of transition rates in J associated with non-arrivals;
(c) Q is the generator matrix of the Markov component J;
(d) $\overline{\sum_p}^{i}$ defines p-th moments of the number of i-calls in one arrival.

The transition probabilities $\pi_{j,k} = P\{J(t) = k | J(0) = j\}$ of the Markov component J form the matrix $\Pi(t) = (\pi_{j,k}(t))$. By that for $h > 0$

$$\pi_{j,k}(h) = \lambda_{j,k}(0)h + o(h), \quad j \neq k,$$

$$\pi_{j,j}(h) = 1 - \gamma_j h + o(h).$$

It is supposed that J is irreducible with stationary distribution $\pi = (\pi_j)$. We note that $\pi\Pi(t) = \pi$, for all t, and $\Pi(t) \to e\pi$, as $t \to \infty$, where $e = (1, 1, ..., 1)^T$ is m-dimensional vector.
According to Corollary 6.1 from Ref. 9, for this stationary distribution π and for the i-th and l-th classes of arrivals, $1 \leq i,\ l \leq r$, the following statements hold.

(a) *Expected values of counts* are calculated as

$$E(X_i(t)) = t\pi \overline{\sum_1}^{i} e. \tag{1}$$

(b) *Variance of the number of counts* $Var(X_i(t))$ satisfies

$$Var(X_i(t))/t \to \pi\Big[\overline{\sum}_2^i + 2\overline{\sum}_1^i \tilde{C} \overline{\sum}_1^i\Big] e, \tag{2}$$

where

$$\widetilde{C} = (I - e\pi)(e\pi - Q)^{-1}. \tag{3}$$

(c) *Covariance of counts* $Cov(X_i(t), X_l(t))$ satisfies

$$Cov(X_i(t), X_l(t))/t \to c_{i,l} \tag{4}$$

where

$$c_{i,l} = \pi\Big[\sum_{\forall n} n_i n_l \Lambda_n + \overline{\sum}_1^i \tilde{C} \overline{\sum}_1^l + \overline{\sum}_1^l \tilde{C} \overline{\sum}_1^i\Big] e. \tag{5}$$

(d) *If*

$$\lambda_i = \pi \overline{\sum}_1^i e < \infty, \forall i,$$

then, for all initial distributions

$$X(t)/t \to \lambda_i \;\; a.s. \tag{6}$$

(e) *If J is finite and*

$$\pi \overline{\sum}_2^i e < \infty, \forall i,$$

then the convergence in distribution to the normal distribution with zero mean vector and covariance matrix $C = (c_{i,j})$ from (c) takes place:

$$\sqrt{t}(X(t)/t - \lambda) \xrightarrow{L} N(0, C).$$

Our paper is devoted to a problem of statistical parameter estimation of the described process. For that we introduce some further simplifications. At first, we pay attention to the *arrival component.* If the state $j \in E$ is fixed, then flows of i-arrivals and l-arrivals, $i \neq l$, are independent. Let n be a scalar value and 0 be a $(r-1)$-dimensional zero vector. Then, let $\lambda_j^1(n) = \lambda_{j,j}\binom{n}{0}$ and $\lambda_j^r(n) = \lambda_{j,j}\binom{0}{n}$ be arrival rates of 1-arrivals and r-arrivals in the state $j \in E$, and analogous notation for i-arrivals, $1 < i < r$. Further, let $q_i(n)$ be a probability that i-arrival contains n items, $\sum_{n>0} q_i(n) = 1$. So, *these probabilities do not depend on state $j \in E$.* Let $\bar{n}_i$ and $\bar{n}_{2,i}$ be *the expected size and the second order moment of the i-batch:*

$$\bar{n}_i = \sum_n n q_i(n), \quad \bar{n}_{2,i} = \sum_n n^2 q_i(n).$$

We suppose that in the state $j \in E$ the *i-arrivals form Poisson flow with rate*

$$\lambda_j^i = \sum_{n>0} \lambda_j^i(n) = v_{j,i}(\alpha^{<i>}), \quad j = 1, ..., m; \quad i = 1, ..., r,$$

where $v_{j,i}$ is *a known function to an approximation of the parameters* $\alpha^{<i>} = (\alpha_{1,i}, \alpha_{2,i}, ..., \alpha_{k,i})^T$.
In particular, for $k = 2$ a linear structure $v_{j,i}(\alpha_{1,i}, \alpha_{2,i}) = \alpha_{1,i} + j\alpha_{2,i}$ is possible.
Now the arrival rates have the following structure:

$$\lambda_j^i(n) = v_{j,i}(\alpha^{<i>})q_i(n), j = 1, ..., m. \tag{7}$$

Secondly, we consider *the Markov component.* We consider a case when J is a birth and death process, with a finite number of states: $\lambda_{j,\xi} = \lambda_{j,\xi}(0) = 0$, if $|j - \xi| > 1$; $j, \xi = 1, ..., m < \infty$. For this case *the stationary distribution of the Markov component* J is calculated as[12] (see Ref. 12, p.193)

$$\pi_1 = \Big(1 + \sum_{j=2}^{m} \prod_{l=1}^{j-1} \frac{\lambda_{l,l+1}}{\lambda_{l+1,l}}\Big)^{-1}, \quad \pi_j = \pi_1 \prod_{l=1}^{j-1} \frac{\lambda_{l,l+1}}{\lambda_{l+1,l}}, \quad j = 2, ..., m. \tag{8}$$

In our case we have the following presentation of the above considered matrices.

(a) Matrix Λ of transition rates in J associated with arrivals is a diagonal matrix:

$$\Lambda = diag\Big(\sum_{i=1}^{r} v_{j,i}(\alpha^{<i>}) : j = 1, ..., m\Big). \tag{9}$$

(b) The generator matrix Q of the Markov component J is of the form:

$$Q = \begin{pmatrix} -\lambda_{1,2} & \lambda_{1,2} & & & \\ \lambda_{2,1} & -(\lambda_{2,1} + \lambda_{2,3}) & \lambda_{2,3} & & \\ & \lambda_{3,2} & -(\lambda_{3,2} + \lambda_{3,4}) & \lambda_{3,4} & \\ \cdots & \cdots & \cdots & \cdots & \cdots \\ & & & \lambda_{m,m-1} & -\lambda_{m,m-1} \end{pmatrix}.$$

(c) The matrices $\overline{\sum}_1^i$ and $\overline{\sum}_2^i$, which take into account the expected size and the second order moments of the size of i-arrivals, are the following:

$$\overline{\sum}_1^i = \bar{n}_i diag(v_{j,i}(\alpha^{<i>}) : j = 1, ..., m),$$

$$\overline{\sum}_2^i = \bar{n}_{2,i} diag(v_{j,i}(\alpha^{<i>}) : j = 1, ..., m). \tag{10}$$

Matrix $\tilde{C}$ is given by Eq. (3).
Now we can calculate indices of interest.

(a) *Expected value of counts is*

$$E(X_i(t))/t = \pi \overline{\sum}_1^i e = \bar{n}_i \sum_{j=1}^{m} v_{j,i}(\alpha^{<i>})\pi_j. \tag{11}$$

(b) *Variance of the number of counts* is of the form

$$Var(X_i(t))/t \approx \pi \left(\overline{\sum}_2^i + 2 \overline{\sum}_1^i \tilde{C} \overline{\sum}_1^i \right) e$$

$$= \bar{n}_{2,i} \sum_{j=1}^{m} \pi_j v_{j,i}(\alpha^{<i>}) + 2\pi \overline{\sum}_1^i \tilde{C} \overline{\sum}_1^i e. \tag{12}$$

(c) The following coefficients are necessary for a *covariance of counts calculation*: for $i \neq l$

$$c_{i,l} = \pi \left[\overline{\sum}_1^i \tilde{C} \overline{\sum}_1^l + \overline{\sum}_1^l \tilde{C} \overline{\sum}_1^i \right] e. \tag{13}$$

Now we are ready for the calculation of these indices.

3. Problem of parameter estimation

Now we consider a problem of unknown parameters estimation. These parameters are the following:

$$\alpha = \left(\alpha^{<1>} \ \alpha^{<2>} \ \ldots \ \alpha^{<r>} \right)_{k \times r}, \quad \overrightarrow{\lambda} = \left(\lambda_{j,j+1} : j = 1, \ldots., m-1 \right),$$

$$\overleftarrow{\lambda} = \left(\lambda_{j,j-1} : j = 2, \ldots., m \right)$$

so $\overrightarrow{\lambda}_j = \lambda_{j,j+1}$, $\overleftarrow{\lambda}_j = \lambda_{j,j-1}$.
We denote them as

$$\theta = (\alpha \ \overrightarrow{\lambda} \ \overleftarrow{\lambda}).$$

It is supposed that we have n independent copies $X^{(1)}(t), \ldots, X^{(n)}(t)$ of the considered process $X(t) = (X_1(t), \ldots, X_r(t))^T$ - total numbers of arrivals of various classes in $(0, t]$. Our initial point, according to the above marked properties, is the following: each $X(t)$ has multivariate normal distribution with mean $E(X(t)) = t\mu$ and covariance matrix $Cov(X(t)) = tC$, where

μ is r-dimensional column-vector and C is $(r \times r)$-matrix. Corresponding empirical mean and covariance matrix

$$\mu^* = \frac{1}{nt}\sum_{i=1}^{n} X^{(i)}(t), \ C^* = \frac{1}{t(n-1)}\sum_{i=1}^{n}\sum_{l=1}^{n}(X^{(i)}(t) - t\mu^*)(X^{(i)}(t) - t\mu^*)^T \tag{14}$$

are sufficient statistics, therefore we must make statistical inferences on this basis.

It is well known from Ref. 13, Theorem 2.5.6, that μ^* has multivariate normal distribution $N_r\left(\mu, \ \frac{1}{nt}C\right)$ and $V = (n-1)tC^*$ has Wishart distribution $W_r(tC, \ n-1)$. If $|C| > 0$ then corresponding density functions are the following:

$$f(\mu^*) = (2\pi)^{-\frac{r}{2}}\left|\frac{1}{nt}C\right|^{-1/2} exp\left\{-\frac{1}{2}(\mu^* - \mu)^T ntC^{-1}(\mu^* - \mu)\right\},$$

$$w(V) = a|tC|^{-\frac{n}{2}}|V|^{(n-r-1)/2} exp\left(-\frac{1}{2t}tr(C^{-1})V\right), \quad V > 0, \tag{15}$$

where a is a normalizing constant and $n > r$.

As μ^* and V are independent, then the log-likelihood function for the given sample is as follows (without normalising constants):

$$l(\alpha, \overrightarrow{\lambda}, \overleftarrow{\lambda}) = ln(f(\mu^*)w(V))$$

$$= -\frac{1}{2}ln(|C|) - \frac{1}{2}(\mu^* - \mu)^T ntC^{-1}(\mu^* - \mu) - \frac{n}{2}ln(|C|) - \frac{1}{2t}tr(C^{-1}V). \tag{16}$$

To find maximum likelihood estimates we must maximize this expression with respect to the unknown parameters. As μ and C are functions of these parameters, μ^* and V are considered as constants. For maximization, a gradient method will be used.

4. Derivatives of the expectation μ

Let us remind some necessary expressions:

$$\mu_i = E(X_i(t))/t = \bar{n}_i\Big(\sum_{j=1}^{m} \pi_j v_{j,i}(\alpha^{<i>})\Big),$$

where

$$\pi_1 = \left(1 + \sum_{j=2}^{m}\prod_{l=1}^{j-1}\frac{\overrightarrow{\lambda}_l}{\overleftarrow{\lambda}_{l+1}}\right)^{-1}, \quad \pi_j = \pi_1\prod_{l=1}^{j-1}\frac{\overrightarrow{\lambda}_l}{\overleftarrow{\lambda}_{l+1}}, \quad j = 2, ..., m.$$

4.1. *Derivatives of* μ *with respect to* α

Let $\nabla\mu_i(\alpha^{<i>})$ be the gradient of μ_i with respect to vector $\alpha^{<i>} = (\alpha_{1,i}\ \ \alpha_{2,i}\ ...\ \alpha_{k,i})$. Then

$$\nabla\mu_i(\alpha^{<i>}) = \Big(\frac{\partial}{\partial\alpha_{1,i}}\mu_i ... \frac{\partial}{\partial\alpha_{k,i}}\mu_i\Big)^T = \bar{n}_i\sum_{j=1}^{m}\pi_j\frac{\partial}{\partial\alpha^{<i>}}v_{j,i}(\alpha^{<i>}),$$

$$\nabla\mu(\alpha) = (\nabla\mu_1(\alpha^{<i>}), ..., \nabla\mu_r(\alpha^{<r>}))_{k\times r}. \tag{17}$$

4.2. *Derivatives of* μ *with respect to* λ

Further, we consider the following column-vectors of dimension $(m-1)$:

$$\nabla\pi_1(\overrightarrow{\lambda}) = \Big(\frac{\partial}{\partial\lambda_{1,2}}\pi_1\ \ ...\ \ \frac{\partial}{\partial\lambda_{m-1,m}}\pi_1\Big)^T$$

$$= -\pi_1^2\Big(\frac{1}{\overrightarrow{\lambda}_1}\sum_{j=2}^{m}\prod_{l=1}^{j-1}\frac{\overrightarrow{\lambda}_l}{\overleftarrow{\lambda}_{l+1}}\ \ \frac{1}{\overrightarrow{\lambda}_2}\sum_{j=3}^{m}\prod_{l=1}^{j-1}\frac{\overrightarrow{\lambda}_l}{\overleftarrow{\lambda}_{l+1}}\ \ ...\ \ \frac{1}{\overrightarrow{\lambda}_{m-1}}\prod_{l=1}^{m-1}\frac{\overrightarrow{\lambda}_l}{\overleftarrow{\lambda}_{l+1}}\Big)^T,$$

$$\nabla\pi_j(\overrightarrow{\lambda}) = \Big(\frac{\partial}{\partial\lambda_{1,2}}\pi_j\ \ ...\ \ \frac{\partial}{\partial\lambda_{m-1,m}}\pi_j\Big)^T = \Big(\prod_{l=1}^{j-1}\frac{\overrightarrow{\lambda}_l}{\overleftarrow{\lambda}_{l+1}}\Big)\nabla\pi_1(\overrightarrow{\lambda})$$

$$+\pi_1\Big(\frac{1}{\overrightarrow{\lambda}_1}\prod_{l=1}^{j-1}\frac{\overrightarrow{\lambda}_l}{\overleftarrow{\lambda}_{l+1}}\ \ \frac{1}{\overrightarrow{\lambda}_2}\prod_{l=1}^{j-1}\frac{\overrightarrow{\lambda}_l}{\overleftarrow{\lambda}_{l+1}}\ \ ...\ \ \frac{1}{\overrightarrow{\lambda}_{j-1}}\prod_{l=1}^{j-1}\frac{\overrightarrow{\lambda}_l}{\overleftarrow{\lambda}_{l+1}}\ 0\ 0\ ...\ 0\Big)^T, \tag{18}$$

for $j = 2, ..., m$.

Other derivatives we find in the similar way:

$$\nabla\pi_1(\overleftarrow{\lambda}) = \Big(\frac{\partial}{\partial\lambda_{2,1}}\pi_1\ \ ...\ \ \frac{\partial}{\partial\lambda_{m,m-1}}\pi_1\Big)^T$$

$$= \pi_1^2\Big(\frac{1}{\overrightarrow{\lambda}_2}\sum_{j=2}^{m}\prod_{l=1}^{j-1}\frac{\overrightarrow{\lambda}_l}{\overleftarrow{\lambda}_{l+1}}\ \ \frac{1}{\overleftarrow{\lambda}_3}\sum_{j=3}^{m}\prod_{l=1}^{j-1}\frac{\overrightarrow{\lambda}_l}{\overleftarrow{\lambda}_{l+1}}\ \ ...\ \ \frac{1}{\overleftarrow{\lambda}_m}\prod_{l=1}^{m-1}\frac{\overrightarrow{\lambda}_l}{\overleftarrow{\lambda}_{l+1}}\Big)^T,$$

$$\nabla\pi_j(\overleftarrow{\lambda}) = \Big(\frac{\partial}{\partial\lambda_{2,1}}\pi_j\ \ ...\ \ \frac{\partial}{\partial\lambda_{m,m-1}}\pi_j\Big) = \Big(\prod_{l=1}^{j-1}\frac{\overrightarrow{\lambda}_l}{\overleftarrow{\lambda}_{l+1}}\Big)\nabla\pi_1(\overleftarrow{\lambda})$$

$$-\pi_1\Big(\frac{1}{\overleftarrow{\lambda}_2}\prod_{l=1}^{j-1}\frac{\overrightarrow{\lambda}_l}{\overleftarrow{\lambda}_{l+1}}\ \frac{1}{\overleftarrow{\lambda}_3}\prod_{l=1}^{j-1}\frac{\overrightarrow{\lambda}_l}{\overleftarrow{\lambda}_{l+1}}\ \ldots\ \frac{1}{\overleftarrow{\lambda}_j}\prod_{l=1}^{j-1}\frac{\overrightarrow{\lambda}_l}{\overleftarrow{\lambda}_{l+1}}\ 0\ 0\ \ldots\ 0\Big)^T, \tag{19}$$

for $j = 2, ..., m$.
Now we can rewrite the obtained results as $(m-1) \times m$-matrices:

$$\nabla\pi(\overrightarrow{\lambda}) = (\nabla\pi_1(\overrightarrow{\lambda})\ \nabla\pi_2(\overrightarrow{\lambda})...\nabla\pi_m(\overrightarrow{\lambda})),$$

$$\nabla\pi(\overleftarrow{\lambda}) = (\nabla\pi_1(\overleftarrow{\lambda})\ \nabla\pi_2(\overleftarrow{\lambda})...\nabla\pi_m(\overleftarrow{\lambda})). \tag{20}$$

Let $A_{<j>}$ denote the j-th row of matrix A, then

$$\frac{\partial}{\partial\lambda_{j,j+1}}\pi = \nabla\pi(\overrightarrow{\lambda})_{<j>}{}^T, \quad \frac{\partial}{\partial\lambda_{j+1,j}}\pi = \nabla\pi(\overleftarrow{\lambda})_{<i>}{}^T. \tag{21}$$

Also for $\mu_i = E(X_i(t))/t = \bar{n}_i \sum_{j=1}^{m} \pi_j v_{j,i}(\alpha^{<i>})$ we have the following vector-functions:

$$\nabla\mu_i(\overrightarrow{\lambda}) = \Big(\frac{\partial}{\partial\overrightarrow{\lambda}_1}\mu_i \ldots \frac{\partial}{\partial\overrightarrow{\lambda}_{m-1}}\mu_i\Big)^T = \bar{n}_i\sum_{j=1}^{m} v_{j,i}(\alpha^{<i>})\nabla\pi_j(\overrightarrow{\lambda}),$$

$$\nabla\mu_i(\overleftarrow{\lambda}) = \Big(\frac{\partial}{\partial\overleftarrow{\lambda}_2}\mu_i \ldots \frac{\partial}{\partial\overleftarrow{\lambda}_m}\mu_i\Big)^T = \bar{n}_i\sum_{j=1}^{m} v_{j,i}(\alpha^{<i>})\nabla\pi_j(\overleftarrow{\lambda}). \tag{22}$$

5. Derivatives of the covariance matrix C

5.1. *Derivatives of C with respect to α*

Let us begin with the following matrices:

$$\overline{\sum}_1^i = \bar{n}_i diag(v_{j,i}(\alpha^{<i>}) : j = 1, ..., m),$$

$$\overline{\sum}_2^i = \bar{n}_{2,i} diag(v_{j,i}(\alpha^{<i>}) : j = 1, ..., m).$$

We have for $p = 1, 2$ setting $\bar{n}_{1,i} = \bar{n}_i$:

$$\begin{aligned}\frac{\partial}{\partial\alpha_{\xi,i}}\overline{\sum}_p^i &= \Big(\frac{\partial}{\partial\alpha_{\xi,i}}\overline{\sum}_p^i\Big)_{m\times m}\\ &= \bar{n}_{p,i} diag\Big(\frac{\partial}{\partial\alpha_{\xi,i}} v_{j,i}(\alpha^{<i>} : j = 1, ..., m)\Big)_{m\times m}.\end{aligned} \tag{23}$$

Now a gradient for the covariance (13), $i \neq l$, can be calculated:

$$\frac{\partial}{\partial \alpha_{\xi,i}} c_{i,l} = \pi^T \Big[\Big(\frac{\partial}{\partial \alpha_{\xi,i}} \overline{\sum}_1^i \Big) \tilde{C} \overline{\sum}_1^l + \overline{\sum}_1^l \tilde{C} \frac{\partial}{\partial \alpha_{\xi,i}} \overline{\sum}_1^i \Big] e$$

$$= \pi^T \Big[\bar{n}_i diag \Big(\frac{\partial}{\partial \alpha_{\xi,i}} v_{j,i}(\alpha^{<i>}) : j = 1, ..., m \Big) \tilde{C} \overline{\sum}_1^l \tag{24}$$

$$+ \overline{\sum}_1^l \tilde{C} \bar{n}_i diag \Big(\frac{\partial}{\partial \alpha_{\xi,i}} v_{j,i}(\alpha^{<i>}) : j = 1, ..., m \Big) \Big] e.$$

If $i = l$ then we must add a term $\bar{n}_{2,i}\pi$ to the previous expression. These derivatives allow us to calculate gradients

$$\nabla c_{i,l}(\alpha^{<i>}) = \Big(\frac{\partial}{\partial \alpha_{1,i}} c_{i,l} ... \frac{\partial}{\partial \alpha_{k,i}} c_{i,l} \Big)^T. \tag{25}$$

5.2. *Derivatives of C with respect to λ*

For the matrix

$$Q = \begin{pmatrix} -\lambda_{1,2} & \lambda_{1,2} & & & \\ \lambda_{2,1} & -(\lambda_{2,1} + \lambda_{2,3}) & \lambda_{2,3} & & \\ \cdots & \cdots & \cdots & \cdots & \cdots \\ & & \lambda_{m-1,m-2} & -(\lambda_{m-1,m-2} + \lambda_{m-1,m}) & \lambda_{m-1,m} \\ & & & \lambda_{m,m-1} & -\lambda_{m,m-1} \end{pmatrix}$$

we get $\frac{\partial}{\partial \lambda_{j,j+1}} Q$ as $m \times m$-matrix with two non-zero elements: 1 as the j-th diagonal element and -1 as the following element on the right-hand side on the j-th row. Matrix $\frac{\partial}{\partial \lambda_{j+1,j}} Q$ has the similar structure but 1 is in the $(j+1)$-th place of the main diagonal and -1 is on its left side on the $(j+1)$-th row.

Now we consider matrix

$$\tilde{C} = (I - e\pi)(e\pi - Q)^{-1}.$$

We get

$$\frac{\partial}{\partial \lambda_{j,j+1}} \tilde{C} = \Big[\frac{\partial}{\partial \lambda_{j,j+1}} (I - e\pi) \Big] (e\pi - Q)^{-1} + (1 - e\pi) \frac{\partial}{\partial \lambda_{j,j+1}} (e\pi - Q)^{-1}$$

$$= -\Big[e \nabla \pi(\vec{\lambda})_{<j>} \Big] (e\pi - Q)^{-1} + (I - e\pi) \frac{\partial}{\partial \lambda_{j,j+1}} (e\pi - Q)^{-1}.$$

Further

$$\frac{\partial}{\partial\lambda_{j,j+1}}(e\pi - Q)^{-1} = (e\pi - Q)^{-1}\Big[\frac{\partial}{\partial\lambda_{j,j+1}}(e\pi - Q)^{-1}\Big](e\pi - Q)^{-1}$$

$$= (e\pi - Q)^{-1}\Big[e\nabla\pi(\overrightarrow{\lambda})_{<j>} - \frac{\partial}{\partial\lambda_{j,j+1}}Q\Big](e\pi - Q)^{-1}.$$

Finally for $j = 1, ..., m-1$

$$\frac{\partial}{\partial\lambda_{j,j+1}}\tilde{C} = \Big\{-\Big[e\nabla\pi(\overrightarrow{\lambda})_{<j>}\Big] - (I - e\pi)(e\pi - Q)^{-1}$$

$$\times\Big[e\nabla\pi(\overrightarrow{\lambda})_{<j>} - \frac{\partial}{\partial\lambda_{j,j+1}}Q\Big]\Big\}(e\pi - Q)^{-1}, \tag{26}$$

$$\frac{\partial}{\partial\lambda_{j+1,j}}\tilde{C} = \Big\{-\Big[e\nabla\pi(\overleftarrow{\lambda})_{<j>}\Big] - (I - e\pi)(e\pi - Q)^{-1}$$

$$\times\Big[e\nabla\pi(\overleftarrow{\lambda})_{<j>} - \frac{\partial}{\partial\lambda_{j+1,j}}Q\Big]\Big\}(e\pi - Q)^{-1}. \tag{27}$$

Now we consider the covariance

$$c_{i,l} = \pi^T\Big[\sum_{\forall n} n_i n_l \Lambda_n + \overline{\sum}_1^i \tilde{C}\overline{\sum}_1^l + \overline{\sum}_1^l \tilde{C}\overline{\sum}_1^i\Big]e.$$

For that we use the matrix derivative technique suggested by Turkington[14]. The corresponding definitions and properties are the following. Let $y = (y_j)$ be an $m \times 1$ vector whose elements are differentiable functions of the elements of an $n \times 1$ vector $x = (x_i)$. Then (see Ref. 14, p.68) *the derivative of y with respect to x*, denoted by $\partial y/\partial x$, is the $n \times m$ matrix given by

$$\frac{\partial y}{\partial x} = \begin{bmatrix} \frac{\partial y_1}{\partial x_1} & \cdots & \frac{\partial y_m}{\partial x_1} \\ \cdots & \cdots & \cdots \\ \frac{\partial y_1}{\partial x_n} & \cdots & \frac{\partial y_m}{\partial x_n} \end{bmatrix}.$$

Note that if y is a scalar, so that $y(x)$ is a scalar function of x, the derivative $\partial y/\partial x$ is the usual gradient, i.e. the $n \times 1$ vector

$$\nabla y(x) = \frac{\partial y}{\partial x} = \Big(\frac{\partial y}{\partial x_1} \cdots \frac{\partial y}{\partial x_n}\Big)^T.$$

We will use *the Chain Rule* of a differentiation (see Ref. 14, p.71) Let $x = (x_i)$, $y = (y_k)$, and $z = (z_j)$ be $n \times 1$, $r \times 1$, and $m \times 1$ vectors, respectively. Suppose z is a vector function of y and y itself is a vector function of x so that $z = z(y(x))$. Then

$$\frac{\partial z}{\partial x} = \frac{\partial y}{\partial x}\frac{\partial z}{\partial y}. \tag{28}$$

The next formulas contain the symbol $\otimes$ of the Kronecker product and the *vec* operator $vec\ M$ that transforms matrix M into a column vector by staking the columns of M one underneath other.

Now let X and A be a nonsingular $n \times n$ matrices and let $|X|$ and trX denote the determinant and trace of X. Then (see Ref. 14, pp. 80, 81)

$$\frac{\partial |X|}{\partial vecX} = |X| vec((X^{-1})^T), \tag{29}$$

$$\frac{\partial vecX^{-1}}{\partial vecX} = -(X^{-1} \otimes (X^{-1})^T), \tag{30}$$

$$\frac{\partial tr\ AX}{\partial vecX} = vecA^T \tag{31}$$

are called *the differentiation rules for determinant, inversion, and trace of matrices.*

We need to perform some auxiliary calculation for vectors a and b and a square matrix M. The chain rule (28) gives

$$\frac{\partial \alpha^T M(\lambda) b}{\partial \lambda} = \frac{\partial vecM(\lambda)}{\partial \lambda}\frac{\partial \alpha^T M b}{\partial vecM} = \frac{\partial vecM(\lambda)}{\partial \lambda}(b \otimes a).$$

Then for $i \neq l$:

$$\begin{aligned}
\nabla c_{i,l}(\vec{\lambda}) &= \nabla \pi^T(\vec{\lambda})\Big(\overline{\sum}_1^i \tilde{C} \overline{\sum}_1^l + \overline{\sum}_1^l \tilde{C} \overline{\sum}_1^i\Big) e \\
&+ \frac{\partial}{\partial \vec{\lambda}} \pi^T \Big(\overline{\sum}_1^i \tilde{C} \overline{\sum}_1^l e + \overline{\sum}_1^l \tilde{C} \overline{\sum}_1^i e\Big)\Big|_{\pi = const} \\
&= \nabla \pi^T(\vec{\lambda})\Big(\overline{\sum}_1^i \tilde{C} \overline{\sum}_1^l + \overline{\sum}_1^l \tilde{C} \overline{\sum}_1^i\Big) e \\
&+ \frac{\partial vec(\tilde{C})}{\partial \vec{\lambda}}\Big[\Big(\overline{\sum}_1^l e \otimes \overline{\sum}_1^i \pi\Big) + \Big(\overline{\sum}_1^i e \otimes \overline{\sum}_1^l \pi\Big)\Big].
\end{aligned} \tag{32}$$

If $i = l$, then we must add to the previous expression a term $\bar{n}_{2,i} \nabla \pi(\vec{\lambda}) v_i(\alpha^{<i>})$ where $v_i(\alpha^{<i>}) = (v_{1,i}(\alpha^{<i>}), ..., v_{m,i}(\alpha^{<i>}))^T$ is the $m \times 1$ vector. For $\overleftarrow{\lambda}$ we have to use the same expression, changing $\vec{\lambda}$ by $\overleftarrow{\lambda}$.

6. Score function

Now we can consider full score function with respect to all unknown parameters $\theta = (\alpha \ \overrightarrow{\lambda} \ \overleftarrow{\lambda})^T$:

$$\nabla l(\theta) = \frac{\partial}{\partial \theta} ln(f(\mu^*)w(V))$$

$$= -\frac{n+1}{2}\frac{\partial}{\partial \theta} ln(|C|) - \frac{nt}{2}\frac{\partial}{\partial \theta}(\mu^* - \mu)^T C^{-1}(\mu^* - \mu) - \frac{1}{2t}\frac{\partial}{\partial \theta} tr(C^{-1}V). \quad (33)$$

Let us find the derivatives by terms. Using the chain rule and the rule for determinants, we have the following presentation for derivative of $ln(|C|)$:

$$\frac{\partial}{\partial \theta} ln(|C|) = \frac{\partial vecC}{\partial \theta}\frac{\partial |C|}{\partial vecC}\frac{\partial ln(|C|)}{\partial |C|} = \frac{\partial vecC}{\partial \theta}|C| vec(C^{-1})\frac{1}{|C|}, \quad (34)$$

where derivatives

$$\frac{\partial vecC}{\partial \theta} = \left(\left(\frac{\partial vecC}{\partial vec\alpha}\right)^T \left(\frac{\partial vecC}{\partial \overrightarrow{\lambda}}\right)^T \left(\frac{\partial vecC}{\partial \overleftarrow{\lambda}}\right)^T\right)^T$$

have been calculated above (see Eqs. (24) - (27)).
Derivatives of a matrix trace are calculated as follows:

$$\frac{\partial}{\partial \theta} tr(C^{-1}V) = \frac{\partial}{\partial \theta} vec(C)\frac{\partial}{\partial vec(C)} vec(C^{-1})\frac{\partial}{\partial vec(C^{-1})} tr(C^{-1}V)$$

$$= \frac{\partial vec(C)}{\partial \theta}\left(-C^{-1} \otimes C^{-1}\right)\frac{\partial}{\partial vec(C^{-1})} tr(VC^{-1})$$

$$= -\frac{\partial vec(C)}{\partial \theta}(C^{-1} \otimes C^{-1})vec(V), \quad (35)$$

where Eqs. (28), (30) and (31) have been used.
Now we find the last derivatives:

$$\frac{\partial}{\partial \theta}(\mu^* - \mu)^T C^{-1}(\mu^* - \mu) = \frac{\partial \mu}{\partial \theta}\frac{\partial}{\partial \mu}(\mu^* - \mu)^T C^{-1}(\mu^* - \mu)$$

$$+ \frac{\partial vecC}{\partial \theta}\frac{\partial vecC^{-1}}{\partial vecC}\frac{\partial}{\partial vecC^{-1}}(\mu^* - \mu)^T C^{-1}(\mu^* - \mu)$$

$$= \frac{\partial \mu}{\partial \theta} 2C^{-1}(\mu^* - \mu) - \frac{\partial vecC}{\partial \theta}(C^{-1} \otimes C^{-1})((\mu^* - \mu) \otimes (\mu^* - \mu))$$

$$= \frac{\partial \mu}{\partial \theta} 2C^{-1}(\mu^* - \mu) - \frac{\partial vecC}{\partial \theta}(C^{-1}(\mu^* - \mu) \otimes C^{-1}(\mu^* - \mu)), \quad (36)$$

where derivatives $\partial vecC/\partial\theta$ and

$$\frac{\partial\mu}{\partial\theta}=\left(\left(\frac{\partial\mu}{\partial\alpha}\right)^T\left(\frac{\partial\mu}{\partial\overrightarrow{\lambda}}\right)^T\left(\frac{\partial\mu}{\partial\overleftarrow{\lambda}}\right)^T\right)^T$$

have been calculated above (see Eqs. (17), (22), (24) - (27)).
Now all terms of the score function (33) are found and we are able to use the gradient method to find maximum likelihood estimates of $\theta=(\alpha\ \ \overrightarrow{\lambda}\ \ \overleftarrow{\lambda})^T$.

7. Parameter identification

Now we shall to discuss a problem of parameter identification. Firstly, it should be noted that above we have supposed that all parameters from $\theta=(\alpha\ \ \overrightarrow{\lambda}\ \ \overleftarrow{\lambda})^T$ are unknown. If some parameters are known, we must set:

(1) zero values for corresponding derivatives,
(2) known parameter values for initial points of our iterative optimization procedure.

Secondly, total number of unknown parameters must not exceed the number of used statistics. In the general case the total number of unknown parameters equals $2(m-1)+k\times r$. We have at our disposal r mean values and $r(r+1)/2$ values of empirical covariances and variances. Therefore, for parameter identification the following inequality has to hold:

$$2(m-1)+k\times r\leq r+r(r+1)/2,$$

$$4(m-1)\leq r(r+3-2k).$$

For example, if $m=k=2$ then $r\geq 3$; if $m=3$, $k=2$ then $r\geq 4$.
Thirdly, indeterminacy (uncertainty) can take place if unknown parameters are estimated up to a scale multiplier. This danger arises in expressions like $\lambda_{j,j+1}/\lambda_{j+1,j}$. But parameters $\lambda_{j,j+1},\lambda_{j+1,j}$ appear in other combinations too. A simple method of verification that the indeterminacy is absent is the following: we use various initial points for our optimization procedure and are satisfied if the obtained results coincide.

8. Numerical example

Our example has the following input data: $m=2, r=3, k=2$, $\alpha^{<i>}=(\alpha_{1,i},\alpha_{i,2})^T$, $v_{j,i}(\alpha^{<i>})=\alpha_{j,i}$, $j=1,2; i=1,2,3$. Therefore $\lambda_j^i=\alpha_{j,i}$ is the rate of i-arrivals in the state $j\in E$. We suppose that 3-calls arrive during the second state and 2-calls arrive during the first state only. Therefore,

$\alpha_{1,3} = \alpha_{2,2} = 0$ and we have 6 unknown parameters $\alpha_{1,1}, \alpha_{1,2}, \alpha_{2,1}, \alpha_{2,3}$; $\overrightarrow{\lambda} = (\lambda_{1,2})$, $\overleftarrow{\lambda} = (\lambda_{2,1})$ and 9 statistical values $\mu^* = (\mu_1^* \; \mu_2^* \; \mu_3^*)$ and $C^* = (c^*_{i,l})_{3\times 3}$.
Probabilistic indices of interest are the following:

$$\pi_1 = \frac{\lambda_{2,1}}{\lambda_{1,2} + \lambda_{2,1}}, \quad \pi_2 = \frac{\lambda_{1,2}}{\lambda_{1,2} + \lambda_{2,1}},$$

$$Q = \begin{pmatrix} -\lambda_{1,2} & \lambda_{1,2} \\ \lambda_{2,1} & -\lambda_{2,1} \end{pmatrix}, \quad \alpha = \begin{pmatrix} \alpha_{1,1} & \alpha_{1,2} & 0 \\ \alpha_{2,1} & 0 & \alpha_{2,3} \end{pmatrix},$$

$$\overline{\sum}_1^i = \bar{n}_i diag(\alpha_{1,i} \; \alpha_{2,i}), \overline{\sum}_2^i = \bar{n}_{2,i} diag(\alpha_{1,i} \; \alpha_{2,i}).$$

$$\tilde{C} = (I - e\pi)(e\pi - Q)^{-1} = \begin{pmatrix} 1-\pi_1 & -1+\pi_1 \\ -\pi_1 & \pi_1 \end{pmatrix} \begin{pmatrix} \pi_1 + \lambda_{1,2} & 1 - \pi_1 - \lambda_{1,2} \\ \pi_1 + \lambda_{2,1} & 1 - \pi_1 + \lambda_{2,1} \end{pmatrix}^{-1}.$$

$$\mu_1 = E(X_i(t))/t \approx \bar{n}_i \Big(\sum_{j=1}^{2} \pi_j \alpha_{j,i}\Big),$$

$$Var(X_i(t))/t \approx \bar{n}_{2,i} \sum_{j=1}^{2} \pi_j \alpha_{j,i} + 2\pi \overline{\sum}_1^i \tilde{C} \overline{\sum}_1^i e.$$

Covariance for $i \neq l$ equals

$$c_{i,l} = \pi \Big[\overline{\sum}_1^i \tilde{C} \overline{\sum}_1^l + \overline{\sum}_1^l \tilde{C} \overline{\sum}_1^i \Big] e.$$

The derivatives have the following form:

$$\frac{\partial}{\partial \alpha_{\xi,i}} v_{j,i}(\alpha^{<i>}) = \frac{\partial}{\partial \alpha_{\xi,i}} (\alpha_{j,i}) = \begin{cases} 1 \; if \;\; \xi = j, \\ 0 \;\; otherwise. \end{cases}$$

$$\Big(\frac{\partial}{\partial \alpha_{\xi,i}} \overline{\sum}_p^i \Big)_{2\times 2} = \bar{n}_{p,i} diag(d(i)),$$

where $d(i)$ is a vector with an unique nonzero component that equals 1 and occupies the i-th place.
The following empirical data are considered: $n = 10, t = 4$,

$$\mu^* = \begin{pmatrix} 1.1 \\ 0.9 \\ 1.3 \end{pmatrix}, \quad C^* = \begin{pmatrix} 0.980 & 0.003 & -0.002 \\ 0.003 & 2.570 & -1.560 \\ -0.002 & -1.560 & 2.740 \end{pmatrix}.$$

Later, we present results of statistical estimation of parameters $\theta = (\alpha_{1,1}\ \alpha_{1,2}\ \alpha_{2,1}\ \alpha_{2,3}\ \lambda_{1,2}\ \lambda_{2,1})^T$. They have been received by the above described procedure for gradient optimization of the log-likelihood function (16).

For initial point $\theta = (1\ 1\ 2\ 2\ 0.5\ 0.7)^T$ the gradient method gives estimates

$$\tilde{\theta} = (1.068\ 1.064\ 2.206\ 2.111\ 0.716\ 1.084)^T$$

with $l(\tilde{\theta}) = -21.823$.

These estimates give the following plug-in estimates for the main indices μ and C:

$$\tilde{\mu} = \begin{pmatrix} 1.066 \\ 0.877 \\ 1.271 \end{pmatrix}, \quad \tilde{C} = \begin{pmatrix} 1.066 & 0.002 & -0.002 \\ 0.002 & 2.173 & -1.239 \\ -0.002 & -1.239 & 2.456 \end{pmatrix}.$$

The same estimates we get for various initial points $\{\theta^{(i)}\}$, which are presented in Table 1. We see that the received estimates have been defined correctly.

Table 1. Considered initial points $\theta^{(i)}$ of optimization procedure.

	$\theta^{(1)}$	$\theta^{(2)}$	$\theta^{(3)}$	$\theta^{(4)}$	$\theta^{(5)}$	$\theta^{(6)}$	$\theta^{(7)}$	$\theta^{(8)}$
$\alpha_{1,1}$	1.0	2.0	0.2	0.0	1.0	0.0	0.0	4.0
$\alpha_{1,2}$	1.0	3.0	0.3	3.0	0.0	1.0	1.0	1.0
$\alpha_{2,1}$	1.0	1.0	0.1	1.0	2.0	1.0	1.0	1.0
$\alpha_{2,3}$	1.0	1.0	0.1	2.1	1.0	2.0	0.2	0.2
$\lambda_{1,2}$	0.6	0.9	0.9	0.9	1.7	0.01	0.9	0.9
$\lambda_{2,1}$	0.6	0.5	0.5	0.5	0.01	1.0	0.9	0.5

9. Conclusions

We have considered a problem of parameter estimation for a Markov-Additive Process of arrivals using maximum likelihood method. A corresponding score function has been derived. Optimization procedure was performed by means of the gradient method. Numerical example shows that the suggested approach gives good results.

Acknowledgement

The author sincerely thanks anonymous reviewers and associate editor, whose remarks have significantly improved this paper.

References

1. F.Hernandez-Campos, K.F.Jeffay, C.Park, J.S Marron and S.I. Resnick, Extremal dependence: Internet traffic applications, *Stochastic Models*, Vol. 22(1), (2005), pp.1-35.
2. D.P.Heyman, and D.Lucantoni, Modelling multiple IP traffic streams with rate limits, *IEE/ACM Transactions on Networking (TON)*. Vol. 11(6), (2003), pp.948-958.
3. C.S.Kim, V.Klimenok, V.Mushko and A.Dudin, The BMAP/PH/N retrial queuing system operating in Markovian random environment, *Computers and Operations Research*, Vol. 37(7), (2010), pp.1228-1237.
4. C.S.Kim, A.Dudin, V.Klimenok and V.Khramova, Erlang loss queuing system with batch arrivals operating in a random environment, *Computers and Operations Research*, Vol. 36(3), (2009), pp.674-697.
5. M.F.Neuts, *Matrix-geometric Solution in Stochastic Models* (The John Hopkins University Press, Baltimore, 1981).
6. M.F.Neuts, *Structured Stochastic Matrices of M/G/1 Type and their Applications* (Marcel Dekker, New York, 1985).
7. S.L.Scott, Bayesian analysis of a two state Markov modulated Poisson process, *J. Comp. Graph. Statist*, Vol. 8, (1999), pp.662-670.
8. S.L.Scott and P.Smyth, The Markov Modulated Poisson Process and Markov Poisson Cascade with Applications to Web Traffic Modelling, *Bayesian Statistics*. Vol. 7, Oxford University Press, (2003), pp.1-10.
9. A.Pacheco, L.C.Tang and N.U.Prabhu, *Markov-Modulated processes and Semiregenerative Phenomena* World Scientific, New Jersey-London-Singapore, (2009).
10. A.C. Davison and N.I Ramesh, Some models for discretised series of events, *J. American Statistical Association*, Vol. 91, (1996), pp.601-609.
11. Yu.Turin, Fitting probabilistic automata via the EM algorithm, *Comm. Statist. Stochastic Models*, Vol. 12, (1996), pp.405-424.
12. M.Kijima, *Markov Processes for Stochastic Modelling* (Chapman & Hall, London, 1997).
13. M.S.Srivastava, *Methods of Multivariate Statistics* (Wiley-Interscience, New York, 2001).
14. D.A.Turkington, *Matrix Calculus and Zero-One Matrices. Statistical and Econometric Applications* (Cambridge University Press, Cambridge, 2002).

A SIMPLE AND EFFICIENT METHOD OF ESTIMATION OF THE PARAMETERS OF A BIVARIATE BIRNBAUM-SAUNDERS DISTRIBUTION BASED ON TYPE-II CENSORED SAMPLES

N. BALAKRISHNAN* and XIAOJUN ZHU**

Department of Mathematics and Statistics, McMaster University, Hamilton, Ontario, Canada L8S 4K1

**E-mail: bala@mcmaster.ca, **E-mail: zhux23@math.mcmaster.ca*

In this paper, we propose a method of estimation for the parameters of a bivariate Birnbaum-Saunders distribution based on Type-II censored samples. The distributional relationship between the bivariate normal and bivariate Birnbaum-Saunders distribution is used for the development of these estimators. The performance of the estimators are then assessed by means of Monte Carlo simulations. Finally, an example is used to illustrate the method of estimation developed here.

Keywords: Bivariate normal distribution; Bivariate Birnbaum-Saunders distribution; Maximum likelihood estimates; Concomitants; Existence; Uniqueness; Correlation; Order statistics; Type-II censoring.

1. Introduction

The Birnbaum-Saunders distribution, proposed originally by Birnbaum and Saunders,[1] is a flexible and useful model for analyzing lifetime data. The cumulative distribution function (CDF) of a two-parameter Birnbaum-Saunders random variable T is given by

$$F(t;\alpha,\beta) = \Phi\left[\frac{1}{\alpha}\left(\sqrt{\frac{t}{\beta}} - \sqrt{\frac{\beta}{t}}\right)\right], \qquad t > 0,\ \alpha > 0,\ \beta > 0, \tag{1}$$

where $\Phi(\cdot)$ is the standard normal CDF.

The Birnbaum-Saunders distribution has found applications in a wide array of problems. For example, Birnbaum and Saunders[2] fitted the model to several data sets on the fatigue life of 6061-T6 aluminum coupons. Desmond[3] extended the model to failure in random environments and investigated the fatigue damage at the root of a cantilever beam. Chang and

Tang[4] applied the model to active repair time for an airborne communication transceiver. For a concise review of developments on Birnbaum-Saunders distribution, one may refer to Johnson, Kotz and Balakrishnan.[5]

Recently, through a transformation of the bivariate normal distribution, Kundu, Balakrishnan and Jamalizadeh[6] derived a bivariate Birnbaum-Saunders distribution, and then discussed the maximum likelihood estimation and modified moment estimation of the five parameters of the model based on complete samples. Here, we consider the situation when the available sample is Type-II censored and develop a method of estimation for the model parameters.

The rest of this paper proceeds as follows. In Section 2, we first describe briefly the bivariate Birnbaum-Saunders distribution and some of its key properties. In Section 3, the nature of Type-II censoring and the form of the available data are presented. In Section 4, we develop the method of estimation for all five parameters of the bivariate Birnbaum-Saunders distribution based on Type-II censored samples. In Section 5, a Monte Carlo simulation study is carried out to examine the bias and mean square error of the proposed estimators for different choices of the parameters and sample sizes. In Section 6, an example is presented to illustrate the proposed method of estimation. The obtained estimates are also compared with the maximum likelihood estimates and the modified moment estimates in the complete sample situation. Finally, in Section 7, we make some concluding remarks and also point out some problems that are of interest for further study.

2. Bivariate Birnbaum-Saunders Distribution and Some Properties

The bivariate random vector (T_1, T_2) is said to have a bivariate Birnbaum-Saunders (BVBS) if it has the joint CDF as

$$P(T_1 \leq t_1, T_2 \leq t_2) = \Phi_2\left[\frac{1}{\alpha_1}\left(\sqrt{\frac{t_1}{\beta_1}} - \sqrt{\frac{\beta_1}{t_1}}\right), \frac{1}{\alpha_2}\left(\sqrt{\frac{t_2}{\beta_2}} - \sqrt{\frac{\beta_2}{t_2}}\right); \rho\right],$$
$$t_1 > 0,\ t_2 > 0, \quad (2)$$

where $\alpha_1 > 0$ and $\alpha_2 > 0$ are the shape parameters, $\beta_1 > 0$ and $\beta_2 > 0$ are the scale parameters, and $-1 < \rho < 1$ is the dependence parameter, and $\Phi_2(z_1, z_2; \rho)$ is the joint CDF of a standard bivariate normal vector (Z_1, Z_2) with correlation coefficient ρ.

Then, the corresponding joint probability density function (PDF) of (T_1, T_2) is given by

$$f_{T_1,T_2}(t_1,t_2) = \phi_2\left[\frac{1}{\alpha_1}\left(\sqrt{\frac{t_1}{\beta_1}} - \sqrt{\frac{\beta_1}{t_1}}\right), \frac{1}{\alpha_2}\left(\sqrt{\frac{t_2}{\beta_2}} - \sqrt{\frac{\beta_2}{t_2}}\right); \rho\right]$$
$$\times \frac{1}{2\alpha_1\beta_1}\left\{\left(\frac{\beta_1}{t_1}\right)^{\frac{1}{2}} + \left(\frac{\beta_1}{t_1}\right)^{\frac{3}{2}}\right\} \frac{1}{2\alpha_2\beta_2}\left\{\left(\frac{\beta_2}{t_2}\right)^{\frac{1}{2}} + \left(\frac{\beta_2}{t_2}\right)^{\frac{3}{2}}\right\},$$
$$t_1 > 0,\ t_2 > 0, \quad (3)$$

where $\phi_2(z_1, z_2; \rho)$ is the joint PDF of Z_1 and Z_2 given by

$$\phi_2(z_1, z_2, \rho) = \frac{1}{2\pi\sqrt{1-\rho^2}} \exp\left[-\frac{1}{2(1-\rho^2)}(z_1^2 + z_2^2 - 2\rho z_1 z_2)\right].$$

Then, the following interesting properties of the bivariate Birnbaum-Saunders distribution in (2) are well-known; see, for example, Kundu, Balakrishnan and Jamalizadeh.[6]

Property 2.1. If $(T_1, T_2) \sim BVBS(\alpha_1, \beta_1, \alpha_2, \beta_2, \rho)$ as defined in (2), then:

- $\frac{1}{\alpha_i}\left(\sqrt{\frac{T_i}{\beta_i}} - \sqrt{\frac{\beta_i}{T_i}}\right) \sim N(0,1)$, $i = 1, 2$;
- $T_i \sim BS(\alpha_i, \beta_i)$, $i = 1, 2$;
- $(T_1^{-1}, T_2^{-1}) \sim BVBS(\alpha_1, \beta_1^{-1}, \alpha_2, \beta_2^{-1}, \rho)$;
- $(T_1^{-1}, T_2) \sim BVBS(\alpha_1, \beta_1^{-1}, \alpha_2, \beta_2, -\rho)$;
- $(T_1, T_2^{-1}) \sim BVBS(\alpha_1, \beta_1, \alpha_2, \beta_2^{-1}, -\rho)$.

3. Form of Data

Suppose (X_i, Y_i), $i = 1, \cdots, n$, denote n independent and identically distributed observations from the bivariate Birnbaum-Saunders distribution in (2). Further, suppose such a complete sample is not available from the n subjects, but only a Type-II censored sample of the following form is observed. Suppose the ordered values of Y are directly observable, and that the first k order statistics of Y-values, denoted by $Y_{1:n} < Y_{2:n} < \cdots < Y_{k:n}$ (for $k \leqslant n$), are observed. For X-variable, the corresponding X-values are observed, which are called the *concomitants of order statistics* [see, for example, David and Nagaraja[7]] and are denoted by $X_{[1:n]}, \cdots, X_{[k:n]}$; that is, we have $X_{[i:n]} = X_j$ if $Y_{i:n} = Y_j$ for $i = 1, \cdots, k$, $j = 1, \cdots, n$. Thus, we assume that the available data from the bivariate Birnbaum-Saunders distribution in (2) is of the form

$$(X_{[1:n]}, Y_{1:n}), (X_{[2:n]}, Y_{2:n}), \cdots, (X_{[k:n]}, Y_{k:n}), \quad (4)$$

where $k(\leqslant n)$ is the pre-fixed number of order statistics from Y-values to be observed. Of course, when $k = n$, we will have a complete sample in an ordered form based on the Y-values.

Based on the form of data in (4), we develop in the next section a simple method of estimation for the parameters α_1, α_2, β_1, β_2 and ρ of the bivariate Birnbaum-Saunders distribution in (2).

4. Estimation Based on Type-II Censored Samples

From (2), it is evident that the bivariate Birnbaum-Saunders distribution has a close relationship with the bivariate normal distribution. From Harrell and Sen,[8] it is known that based on a Type-II censored data of the form (4), the maximum likelihood estimates of the parameters of the bivariate normal distribution are given by

$$\hat{\mu}_X = \bar{X} + (S_{XY}/S_Y^2)(\hat{\mu}_Y - \bar{Y}), \tag{5}$$
$$\hat{\sigma}_X^2 = S_X^2 + (S_{XY}^2/S_Y^2)(\hat{\sigma}_Y^2/S_Y^2 - 1), \tag{6}$$
$$\hat{\rho} = \hat{\sigma}_Y S_{XY}/(\hat{\sigma}_X S_Y^2), \tag{7}$$

where,

$$\bar{X} = \frac{1}{k}\sum_{i=1}^{k} X_{[i:n]}, \qquad \bar{Y} = \frac{1}{k}\sum_{i=1}^{k} Y_{i:n}, \qquad S_X^2 = \frac{1}{k}\sum_{i=1}^{k}(X_{[i:n]} - \bar{X})^2,$$
$$S_Y^2 = \frac{1}{k}\sum_{i=1}^{k}(Y_{i:n} - \bar{Y})^2, \qquad S_{XY} = \frac{1}{k}\sum_{i=1}^{k}(X_{[i:n]} - \bar{X})(Y_{i:n} - \bar{Y}); \tag{8}$$

in the above, $\hat{\mu}_Y$ and $\hat{\sigma}_Y$ need to be obtained by solving the corresponding likelihood equations. For more details, one may refer to Harrell and Sen.[8]

Now, we use these results known for the bivariate normal case to develop a simple and efficient method of estimation for the parameters of the bivariate Birnbaum-Saunders distribution in (2). For this, we first of all assume that the censoring present in the data is light or moderate and not heavy, with $\frac{n}{2} < k \leqslant n$. This is necessary for the ensuing method of estimation. Then, we proceed as follows. We first estimate the parameters μ_Y and σ_Y directly from the Type-II censored sample on Y, and then use a transformation [based on (2)] to convert the problem to that of a bivariate normal distribution and determine the estimates of the remaining parameters μ_X, σ_X and ρ.

To this end, let us first of all consider the Type-II right censored sample on Y given by $Y_{1:n}, Y_{2:n}, \cdots, Y_{k:n}$. From Property 2.1, it is known that

$$Y \sim BS(\alpha_2, \beta_2) \qquad and \qquad \frac{1}{Y} \sim BS(\alpha_2, \frac{1}{\beta_2}). \tag{9}$$

Consequently, the order statistics from Y-values satisfy a "reciprocal property", and for utilizing it, we can only make use of the order statistics $Y_{n-k+1:n}, Y_{n-k+2:n}, \cdots, Y_{k:n}$. Then, we propose the following intuitive estimates of β_2 and α_2:

$$\hat{\beta}_2 = \frac{\sum_{i=n-k+1}^{k} \sqrt{Y_{i:n}}}{\sum_{i=n-k+1}^{k} \sqrt{1/Y_{i:n}}} \tag{10}$$

and

$$\hat{\alpha}_2 = \sqrt{\frac{1}{n-1}\left[\sum_{i=1}^{k}\left(\frac{Y_{i:n}}{\hat{\beta}_2} + \frac{\hat{\beta}_2}{Y_{i:n}} - 2\right) + \sum_{i=1}^{n-k}\left(\frac{Y_{i:n}}{\hat{\beta}_2} + \frac{\hat{\beta}_2}{Y_{i:n}} - 2\right)\right]}. \tag{11}$$

Next, for the estimation of α_1, β_1 and ρ, from (2), we consider the transformation

$$Z_X = \sqrt{\frac{X}{\beta_1}} - \sqrt{\frac{\beta_1}{X}} \quad \text{and} \quad Z_Y = \sqrt{\frac{Y}{\beta_2}} - \sqrt{\frac{\beta_2}{Y}}$$

which results in $(Z_X, Z_Y) \sim N(\mathbf{0}, \mathbf{\Sigma})$, where $\Sigma = \begin{bmatrix} \alpha_1 & \rho\alpha_1\alpha_2 \\ \rho\alpha_1\alpha_2 & \alpha_2 \end{bmatrix}$. Now, by using the maximum likelihood estimates of the parameters of the bivariate normal distribution presented earlier, β_1 can be estimated from the following equation

$$\bar{Z}_X = (S_{Z_X Z_Y}/S^2_{Z_Y})\bar{Z}_Y, \tag{12}$$

where,

$$Z_X = \sqrt{\frac{X}{\hat{\beta}_1}} - \sqrt{\frac{\hat{\beta}_1}{X}}, \quad \bar{Z}_X = \frac{1}{k}\sum_{i=1}^{k} Z_{X_{[i:n]}}, \quad S^2_{Z_X} = \frac{1}{k}\sum_{i=1}^{k}(Z_{X_{[i:n]}} - \bar{Z}_X)^2,$$

$$Z_Y = \sqrt{\frac{Y}{\hat{\beta}_2}} - \sqrt{\frac{\hat{\beta}_2}{Y}}, \quad \bar{Z}_Y = \frac{1}{k}\sum_{i=1}^{k} Z_{Y_{i:n}}, \quad S^2_{Z_Y} = \frac{1}{k}\sum_{i=1}^{k}(Z_{Y_{i:n}} - \bar{Z}_Y)^2,$$

$$S_{Z_X Z_Y} = \frac{1}{k}\sum_{i=1}^{k}(Z_{X_{[i:n]}} - \bar{Z}_X)(Z_{Y_{i:n}} - \bar{Z}_Y). \tag{13}$$

Finally, the estimates of $\hat{\alpha}_1$ and $\hat{\rho}$ can be determined as

$$\hat{\alpha}_1 = \sqrt{S^2_{Z_X} + (S^2_{Z_X Z_Y}/S^2_{Z_Y})(\hat{\alpha}_2^2/S^2_{Z_Y} - 1)}, \tag{14}$$

$$\hat{\rho} = \hat{\alpha}_2 S_{Z_X Z_Y}/(\hat{\alpha}_1 S^2_{Z_Y}). \tag{15}$$

The implicit form of the solution for $\hat{\beta}_1$ in (12) makes it difficult to establish the existence and uniqueness of the solution. However, in our extensive Monte Carlo simulations, we have observed that Eq. (12) always resulted in a unique solution for $\hat{\beta}_1$. It is important to mention here that the method is applicable in the complete sample as well.

5. Simulation Study

We carried out an extensive Monte Carlo simulation study for different sample sizes n and values of ρ, by taking $\alpha_1 = 0.25$, $\alpha_2 = 1$, and $\beta_1 = \beta_2 = 1$. We chose sample sizes n to be 20 and 100, and the values of ρ to be 0.95, 0.50, 0.25 and 0.00. With these choices of n and all the parameters, we simulated bivariate Type-II censored data of the form in (4) with the first k order statistics on Y and the corresponding concomitants on X being observed. Then, by using the method of estimation proposed in the preceding Section, we determined the empirical values of the means and mean square errors (MSEs) of the estimates of α_1, α_2, β_1, β_2 and ρ for various degrees of censoring, namely, 0%, 10%, 20%, 30% and 40%. These results are all presented in Tables 1–4. From these tables, we observe that the precision of the estimates do not seem to depend on the value of ρ. Moreover, we observe that all the bias values are very small even when the sample size is as small as 20 which reveals that the proposed estimates are very nearly unbiased.

6. Illustrative Data Analysis

In this Section, we will illustrate the proposed method of estimation by considering a data, used earlier by Kundu, Balakrishnan and Jamalizadeh.[6]

Example 6.1. These data, presented in Table 5, given by Johnson and Wichern,[9] represent the bone mineral density (BMD) measured in g/cm^2 for 24 individuals, who had participated in an experimental study. In this table, the first number represents the BMD of the bone dominant radius before starting the study while the second number represents the BMD of the same bone after one year. Kundu, Balakrishnan and Jamalizadeh[6] analyzed

Table 1. Simulated values of means and MSE (reported within brackets) of the proposed estimates when $\alpha_1 = \alpha_2 = 0.25$, $\beta_1 = \beta_2 = 1$ and $n = 20$. Here, d.o.c. denotes degree of censoring.

ρ	d.o.c.(%)	$\hat{\alpha}_1$	$\hat{\alpha}_2$	$\hat{\beta}_1$	$\hat{\beta}_2$	$\hat{\rho}$
0.95	0	0.2474	0.2474	1.0014	1.0016	0.9504
		(0.0016)	(0.0016)	(0.0032)	(0.0032)	(0.0007)
	10	0.2457	0.2465	1.0012	1.0016	0.9482
		(0.0022)	(0.0022)	(0.0033)	(0.0034)	(0.0009)
	20	0.2453	0.2460	1.0013	1.0017	0.9472
		(0.0027)	(0.0026)	(0.0036)	(0.0036)	(0.0012)
	30	0.2450	0.2461	1.0014	1.0020	0.9463
		(0.0033)	(0.0030)	(0.0040)	(0.0039)	(0.0016)
	40	0.2449	0.2463	1.0018	1.0024	0.9451
		(0.0039)	(0.0035)	(0.0048)	(0.0043)	(0.0022)
0.50	0	0.2474	0.2474	1.0011	1.0019	0.4908
		(0.0016)	(0.0016)	(0.0031)	(0.0033)	(0.0321)
	10	0.2436	0.2467	1.0010	1.0020	0.4869
		(0.0020)	(0.0022)	(0.0034)	(0.0034)	(0.0497)
	20	0.2437	0.2464	1.0009	1.0023	0.4768
		(0.0025)	(0.0027)	(0.0043)	(0.0037)	(0.0663)
	30	0.2457	0.2467	1.0029	1.0030	0.4702
		(0.0032)	(0.0032)	(0.0060)	(0.0041)	(0.0873)
	40	0.2477	0.2470	1.0045	1.0034	0.4582
		(0.0040)	(0.0036)	(0.0089)	(0.0046)	(0.1173)
0.25	0	0.2473	0.2474	1.0010	1.0020	0.2449
		(0.0016)	(0.0016)	(0.0031)	(0.0033)	(0.0477)
	10	0.2435	0.2466	1.0014	1.0021	0.2440
		(0.0019)	(0.0022)	(0.0035)	(0.0034)	(0.0709)
	20	0.2446	0.2465	1.0014	1.0025	0.2349
		(0.0023)	(0.0027)	(0.0046)	(0.0037)	(0.0973)
	30	0.2460	0.2467	1.0025	1.0031	0.2296
		(0.0028)	(0.0032)	(0.0064)	(0.0041)	(0.1201)
	40	0.2491	0.2470	1.0051	1.0035	0.2269
		(0.0035)	(0.0036)	(0.0102)	(0.0046)	(0.1564)
0.00	0	0.2473	0.2474	1.0010	1.0020	0.0012
		(0.0017)	(0.0016)	(0.0031)	(0.0033)	(0.0533)
	10	0.2432	0.2466	1.0006	1.0021	-0.0022
		(0.0019)	(0.0022)	(0.0035)	(0.0034)	(0.0790)
	20	0.2444	0.2465	1.0017	1.0026	0.0011
		(0.0021)	(0.0027)	(0.0045)	(0.0037)	(0.1041)
	30	0.2475	0.2469	1.0031	1.0033	0.0026
		(0.0026)	(0.0032)	(0.0067)	(0.0041)	(0.1343)
	40	0.2505	0.2471	1.0041	1.0037	0.0016
		(0.0033)	(0.0036)	(0.0106)	(0.0046)	(0.1736)

Table 2. Simulated values of means and MSE (reported within brackets) of the proposed estimates when $\alpha_1 = \alpha_2 = 0.25$, $\beta_1 = \beta_2 = 1$ and $n = 100$. Here, d.o.c. denotes degree of censoring.

ρ	d.o.c.(%)	$\hat{\alpha}_1$	$\hat{\alpha}_2$	$\hat{\beta}_1$	$\hat{\beta}_2$	$\hat{\rho}$
0.95	0	0.2500	0.2495	1.0005	1.0007	0.9496
		(0.0003)	(0.0003)	(0.0006)	(0.0006)	(0.0001)
	10	0.2497	0.2496	1.0007	1.0009	0.9499
		(0.0005)	(0.0005)	(0.0007)	(0.0007)	(0.0001)
	20	0.2499	0.2499	1.0010	1.0012	0.9497
		(0.0006)	(0.0005)	(0.0007)	(0.0007)	(0.0002)
	30	0.2499	0.2500	1.0010	1.0014	0.9496
		(0.0007)	(0.0006)	(0.0008)	(0.0008)	(0.0002)
	40	0.2501	0.2502	1.0013	1.0016	0.9495
		(0.0008)	(0.0007)	(0.0010)	(0.0009)	(0.0002)
0.50	0	0.2500	0.2491	1.0003	1.0009	0.4988
		(0.0003)	(0.0003)	(0.0006)	(0.0006)	(0.0057)
	10	0.2495	0.2493	1.0006	1.0010	0.4995
		(0.0004)	(0.0004)	(0.0007)	(0.0007)	(0.0087)
	20	0.2498	0.2496	1.0010	1.0014	0.4983
		(0.0005)	(0.0005)	(0.0008)	(0.0007)	(0.0120)
	30	0.2498	0.2498	1.0011	1.0016	0.4959
		(0.0006)	(0.0006)	(0.0012)	(0.0008)	(0.0155)
	40	0.2503	0.2499	1.0013	1.0018	0.4920
		(0.0008)	(0.0007)	(0.0018)	(0.0009)	(0.0214)
0.25	0	0.2500	0.2491	1.0002	1.0009	0.2497
		(0.0003)	(0.0003)	(0.0006)	(0.0006)	(0.0089)
	10	0.2492	0.2492	1.0004	1.0010	0.2502
		(0.0004)	(0.0004)	(0.0007)	(0.0007)	(0.0139)
	20	0.2500	0.2494	1.0012	1.0014	0.2515
		(0.0004)	(0.0005)	(0.0009)	(0.0007)	(0.0189)
	30	0.2501	0.2496	1.0011	1.0016	0.2488
		(0.0005)	(0.0006)	(0.0013)	(0.0008)	(0.0249)
	40	0.2508	0.2497	1.0015	1.0017	0.2466
		(0.0006)	(0.0007)	(0.0020)	(0.0009)	(0.0331)
0.00	0	0.2500	0.2490	1.0001	1.0009	0.0009
		(0.0003)	(0.0003)	(0.0006)	(0.0006)	(0.0100)
	10	0.2492	0.2491	1.0004	1.0010	0.0024
		(0.0003)	(0.0004)	(0.0007)	(0.0007)	(0.0160)
	20	0.2496	0.2493	1.0008	1.0013	0.0035
		(0.0004)	(0.0005)	(0.0009)	(0.0007)	(0.0217)
	30	0.2503	0.2495	1.0018	1.0015	0.0072
		(0.0004)	(0.0006)	(0.0013)	(0.0008)	(0.0285)
	40	0.2510	0.2495	1.0010	1.0016	0.0029
		(0.0005)	(0.0007)	(0.0020)	(0.0009)	(0.0378)

Table 3. Simulated values of means and MSE (reported within brackets) of the proposed estimates when $\alpha_1 = 0.25$, $\alpha_2 = 1.00$, $\beta_1 = \beta_2 = 1$ and $n = 20$. Here, d.o.c. denotes degree of censoring.

ρ	d.o.c.(%)	$\hat{\alpha}_1$	$\hat{\alpha}_2$	$\hat{\beta}_1$	$\hat{\beta}_2$	$\hat{\rho}$
0.95	0	0.2474	0.9861	1.0014	1.0216	0.9472
		(0.0016)	(0.0257)	(0.0032)	(0.0457)	(0.0007)
	10	0.2458	0.9840	1.0013	1.0242	0.9481
		(0.0022)	(0.0367)	(0.0034)	(0.0521)	(0.0009)
	20	0.2453	0.9829	1.0013	1.0267	0.9471
		(0.0027)	(0.0457)	(0.0037)	(0.0584)	(0.0012)
	30	0.2449	0.9840	1.0013	1.0303	0.9462
		(0.0033)	(0.0540)	(0.0040)	(0.0651)	(0.0016)
	40	0.2447	0.9860	1.0014	1.0348	0.9451
		(0.0039)	(0.0636)	(0.0046)	(0.0740)	(0.0022)
0.50	0	0.2474	0.9861	1.0011	1.0227	0.4907
		(0.0016)	(0.0260)	(0.0031)	(0.0461)	(0.0321)
	10	0.2436	0.9848	1.0010	1.0260	0.4870
		(0.0020)	(0.0373)	(0.0035)	(0.0531)	(0.0497)
	20	0.2438	0.9849	1.0009	1.0298	0.4769
		(0.0025)	(0.0471)	(0.0043)	(0.0602)	(0.0663)
	30	0.2456	0.9872	1.0029	1.0353	0.4702
		(0.0032)	(0.0571)	(0.0060)	(0.0692)	(0.0873)
	40	0.2476	0.9893	1.0044	1.0406	0.4580
		(0.0040)	(0.0674)	(0.0089)	(0.0801)	(0.1173)
0.25	0	0.2473	0.9861	1.0010	1.0230	0.2449
		(0.0016)	(0.0262)	(0.0031)	(0.0461)	(0.0477)
	10	0.2435	0.9845	1.0014	1.0261	0.2441
		(0.0019)	(0.0375)	(0.0035)	(0.0532)	(0.0709)
	20	0.2446	0.9852	1.0015	1.0305	0.2351
		(0.0023)	(0.0476)	(0.0046)	(0.0608)	(0.0973)
	30	0.2460	0.9875	1.0025	1.0360	0.2296
		(0.0028)	(0.0577)	(0.0064)	(0.0702)	(0.1200)
	40	0.2490	0.9897	1.0051	1.0411	0.2269
		(0.0035)	(0.0679)	(0.0101)	(0.0807)	(0.1562)
0.00	0	0.2473	0.9861	1.0010	1.0232	0.0012
		(0.0017)	(0.0264)	(0.0031)	(0.0461)	(0.0532)
	10	0.2432	0.9845	1.0006	1.0263	-0.0022
		(0.0019)	(0.0376)	(0.0035)	(0.0533)	(0.0790)
	20	0.2444	0.9853	1.0017	1.0309	0.0013
		(0.0021)	(0.0479)	(0.0045)	(0.0613)	(0.1040)
	30	0.2475	0.9881	1.0031	1.0367	0.0028
		(0.0026)	(0.0579)	(0.0067)	(0.0706)	(0.1342)
	40	0.2505	0.9901	1.0042	1.0416	0.0017
		(0.0033)	(0.0679)	(0.0106)	(0.0808)	(0.1734)

Table 4. Simulated values of means and MSE (reported within brackets) of the proposed estimates when $\alpha_1 = 0.25$, $\alpha_2 = 1.00$, $\beta_1 = \beta_2 = 1$ and $n = 100$. Here, d.o.c. denotes degree of censoring.

ρ	d.o.c.(%)	$\hat{\alpha}_1$	$\hat{\alpha}_2$	$\hat{\beta}_1$	$\hat{\beta}_2$	$\hat{\rho}$
0.95	0	0.2498	0.9972	1.0005	1.0055	0.9495
		(0.0003)	(0.0050)	(0.0006)	(0.0084)	(0.0001)
	10	0.2498	0.9984	1.0007	1.0070	0.9498
		(0.0005)	(0.0075)	(0.0007)	(0.0098)	(0.0001)
	20	0.2499	1.0001	1.0010	1.0090	0.9497
		(0.0006)	(0.0095)	(0.0007)	(0.0112)	(0.0002)
	30	0.2498	1.0009	1.0010	1.0103	0.9496
		(0.0007)	(0.0113)	(0.0008)	(0.0126)	(0.0002)
	40	0.2500	1.0018	1.0012	1.0116	0.9495
		(0.0008)	(0.0131)	(0.0009)	(0.0142)	(0.0002)
0.50	0	0.2500	0.9959	1.0003	1.0061	0.4987
		(0.0004)	(0.0051)	(0.0006)	(0.0086)	(0.0051)
	10	0.2495	0.9971	1.0006	1.0076	0.4995
		(0.0004)	(0.0073)	(0.0007)	(0.0101)	(0.0087)
	20	0.2498	0.9988	1.0010	1.0096	0.4982
		(0.0005)	(0.0091)	(0.0009)	(0.0115)	(0.0120)
	30	0.2498	0.9998	1.0010	1.0112	0.4958
		(0.0006)	(0.0108)	(0.0012)	(0.0130)	(0.0155)
	40	0.2502	1.0007	1.0012	1.0125	0.4918
		(0.0008)	(0.0124)	(0.0018)	(0.0145)	(0.0214)
0.25	0	0.2500	0.9956	1.0002	1.0063	0.2496
		(0.0003)	(0.0051)	(0.0006)	(0.0086)	(0.0089)
	10	0.2492	0.9966	1.0004	1.0077	0.2502
		(0.0004)	(0.0073)	(0.0007)	(0.0102)	(0.0139)
	20	0.2500	0.9981	1.0012	1.0095	0.2514
		(0.0004)	(0.0091)	(0.0009)	(0.0116)	(0.0189)
	30	0.2501	0.9993	1.0011	1.0112	0.2487
		(0.0005)	(0.0107)	(0.0013)	(0.0131)	(0.0249)
	40	0.2508	0.9997	1.0015	1.0121	0.2465
		(0.0006)	(0.0123)	(0.0020)	(0.0146)	(0.0331)
0.00	0	0.2500	0.9955	1.0001	1.0064	0.0009
		(0.0003)	(0.0051)	(0.0006)	(0.0087)	(0.0101)
	10	0.2492	0.9962	1.0004	1.0076	0.0024
		(0.0003)	(0.0073)	(0.0007)	(0.0103)	(0.0160)
	20	0.2496	0.9977	1.0008	1.0094	0.0035
		(0.0004)	(0.0091)	(0.0009)	(0.0117)	(0.0217)
	30	0.2503	0.9985	1.0018	1.0108	0.0071
		(0.0004)	(0.0107)	(0.0013)	(0.0132)	(0.0285)
	40	0.2510	0.9990	1.0010	1.0119	0.0029
		(0.0005)	(0.0125)	(0.0020)	(0.0149)	(0.0378)

this example by assuming a bivariate Birnbaum-Saunders distribution for the given complete bivariate data.

First, based on the complete bivariate data in Table 5, we determined the maximum likelihood estimates (MLEs), modified moment estimates (MMEs), and the proposed estimates of the parameters α_1, α_2, β_1, β_2 and ρ. These results are all presented in Table 6 from which we immediately observe that the proposed estimates are indeed very nearly the same as the other two sets of estimates in this complete sample situation. Incidentally, the values of MLEs and MMEs in this case have also been reported earlier by Kundu, Balakrishnan and Jamalizadeh.[6]

Next, with the estimates obtained by these three methods, we also carried out the Kolmogorov-Smirnov test for the goodness-of-fit of the Birnbaum-Saunders distribution for the data on individual components X and Y in Table 5. The computed values of the Kolmogorov-Smirnov distances and the corresponding P-values for all three methods of estimation are presented in Table 7, and these results do support the model assumption made in our analysis.

Finally, for the purpose of illustrating the proposed method of estimation for Type-II censored data, we introduced various levels of censoring in the data presented in Table 5 by taking $k = 22$, 20, 18 and 16. We then determined the estimates of the parameters α_1, α_2, β_1, β_2 and ρ based on these censored data, and these are all presented in Table 8. Their standard errors were determined by Monte Carlo simulations, and these values also reported in Table 8. Upon comparing the estimates of the parameters in Table 8 with the corresponding ones in Table 6 based on the complete data, we observe that the estimates of the parameters are all close even in the case of $k = 16$ (i.e., 8 of the 24 observations censored). We also note that, even though the estimates remain fairly stable when amount of censoring increases, the standard errors of the estimates do increase, as one would expect. Finally, in Figure 1, we have demonstrated the existence and uniqueness of the estimate of β_1 when solving the corresponding estimating equation.

7. Concluding Remarks

In this paper, we have developed a simple and efficient method of estimating the five parameters of the bivariate Birnbaum-Saunders distribution in (2) when the available data are Type-II censored and are of the form in (4). In developing this method of estimation, we have assumed that the censoring present in the data is light or moderate, with $\frac{n}{2} < k \leqslant n$. This is

Table 5. The bone mineral density data taken from Johnson and Wichern.[9]

X	1.103	0.842	0.925	0.857	0.795	0.787
Y	1.027	0.857	0.875	0.873	0.811	0.640
X	0.933	0.799	0.945	0.921	0.792	0.815
Y	0.947	0.886	0.991	0.977	0.825	0.851
X	0.755	0.880	0.900	0.764	0.733	0.932
Y	0.770	0.912	0.905	0.756	0.765	0.932
X	0.856	0.890	0.688	0.940	0.493	0.835
Y	0.843	0.879	0.673	0.949	0.463	0.776

Table 6. Estimates of the parameters based on data in Table 5.

Estimation	$\hat{\alpha}_1$	$\hat{\beta}_1$	$\hat{\alpha}_2$	$\hat{\beta}_2$	$\hat{\rho}$
MMEs	0.1491	0.8316	0.1674	0.8293	0.9343
MLEs	0.1491	0.8313	0.1674	0.8292	0.9343
Proposed estimates	0.1491	0.8321	0.1674	0.8302	0.9343

Table 7. Kolmogorov-Smirnov distance and the corresponding P-value for the BS goodness-of-fit for the data on components X and Y in Table 5.

	X		Y	
Estimation	Distance	P-value	Distance	P-value
MMEs	0.1537	0.5696	0.1650	0.4805
MLEs	0.1530	0.5757	0.1654	0.4777
Proposed estimates	0.1549	0.5599	0.1629	0.4969

necessary for the method of estimation as it uses the reciprocal property of the Birnbaum-Saunders distribution. This leaves some problems open for further study. First, in the present Type-II censoring situation considered here, it would be of interest to develop a method of estimation when the censoring present is heavy, i.e., $k \leqslant \frac{n}{2}$. Next, it would also be of interest to extend the proposed method of estimation to the case when the available data are Type-I censored and progressively Type-II censored; interested readers may refer to Balakrishnan and Aggarwala[10] and Balakrishnan[11] for a detailed review of various developments on progressive censoring. With regard to the latter, the results of Balakrishnan and Kim,[12] extending the work of Harrell and Sen[8] to the case of progressive Type-II censoring, would prove to be quite useful.

Table 8. Estimates of the parameters and standard errors (reported within brackets) based on Type-II censored data on Y, where k denotes the rank of the last observed order statistic from Y.

k	$\hat{\alpha}_1$	$\hat{\beta}_1$	$\hat{\alpha}_2$	$\hat{\beta}_2$	$\hat{\rho}$
22	0.1808	0.8438	0.2175	0.8496	0.9589
	(0.0302)	(0.0323)	(0.0364)	(0.0393)	(0.0206)
20	0.1871	0.8484	0.2242	0.8550	0.9583
	(0.0347)	(0.0347)	(0.0419)	(0.0409)	(0.0240)
18	0.1861	0.8471	0.2260	0.8569	0.9537
	(0.0375)	(0.0359)	(0.0442)	(0.0437)	(0.0303)
16	0.1891	0.8496	0.2296	0.8609	0.9464
	(0.0414)	(0.0389)	(0.0475)	(0.0463)	(0.0463)

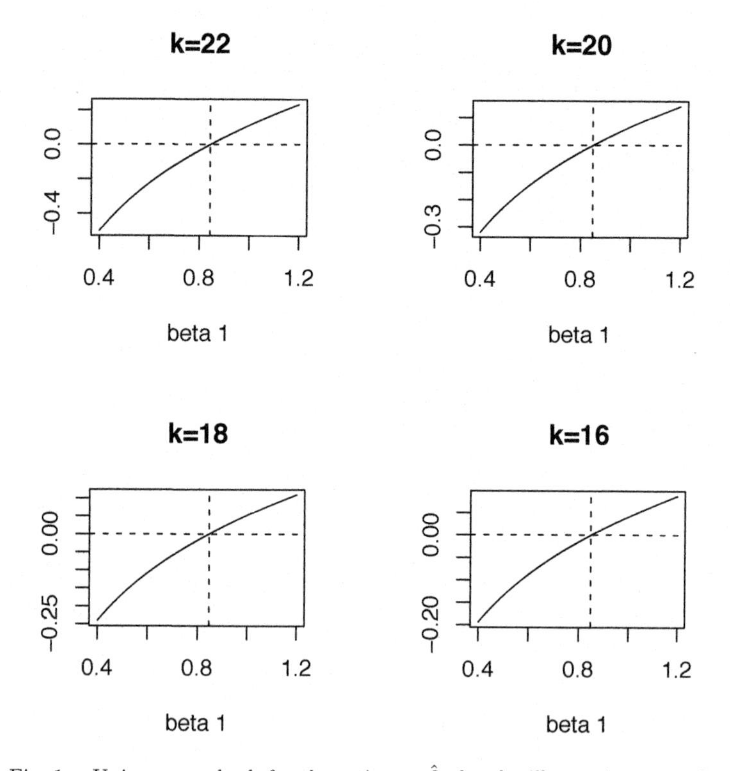

Fig. 1. Uniqueness check for the estimate $\hat{\beta}_1$ for the illustrative example.

Acknowledgments. The first author thanks the Natural Sciences and Engineering Research Council of Canada for funding this research work. The authors also thank the editors of this volume for extending an invitation to prepare this article, and the anonymous reviewers for making some suggestions on an earlier version of this manuscript which led to this improved version.

References

1. Z. W. Birnbaum and S. C. Saunders, A new family of life distributions, *Journal of Applied Probability* **6**, 319–327 (1969).
2. Z. W. Birnbaum and S. C. Saunders, Estimation for a family of life distributions with applications to fatigue, *Journal of Applied Probability* **6**, 328–347 (1969).
3. A. F. Desmond, Stochastic models of failure in random environments, *Canadian Journal of Statistics* **13**, 171–183 (1985).
4. D. S. Chang and L. C. Tang, Reliability bounds and critical time for the Birnbaum-Saunders distribution, *IEEE Transactions on Reliability* **42**, 464–469 (1993).
5. N. L. Johnson, S. Kotz and N. Balakrishnan, *Continuous Univariate Distributions–Vol. 2*, Second edition (John Wiley & Sons, New York 1995).
6. D. Kundu, N. Balakrishnan and A. Jamalizadeh, Bivariate Birnbaum-Saunders distribution and associated inference, *Journal of Multivariate Analysis* **101**, 113–125 (2010).
7. H. A. David and H. N. Nagaraja, *Order Statistics*, Third edition (John Wiley & Sons, Hoboken, New Jersey 2003).
8. F. E. Harrell and P. K. Sen, Statistical inference for censored bivariate normal distribution based on induced order statistics, *Biometrika* **66**, 293–298 (1979).
9. R. A. Johnson and D. W. Wichern, *Applied Multivariate Analysis*, Fourth edition (Prentice-Hall, New Jersey 1999).
10. N. Balakrishnan and R. Aggarwala, *Progressive Censoring: Theory, Methods, and Applications* (Birkhäuser, Boston 2000).
11. N. Balakrishnan, Progressive censoring methodology: An appraisal (with discussions), *Test* **16**, 211–296 (2007).
12. N. Balakrishnan and J.-A. Kim, *EM algorithm and optimal censoring schemes for progressively Type-II censored bivariate normal data*, in *Advances in Ranking and Selection, Multiple Comparisons, and Reliability* eds. N. Balakrishnan, N. Kannan and H.N. Nagaraja (Birkhäuser, Boston 2005), pp. 21–45.

ANALYSIS OF CONTINGENT VALUATION DATA WITH SELF-SELECTED ROUNDED WTP-INTERVALS COLLECTED BY TWO-STEPS SAMPLING PLANS

YU. K. BELYAEV* and B. KRISTRÖM

Department of Mathematics and Mathematical Statistics, Umeå University, Umeå, SE-90 187, Sweden
**E-mail: yuri.belyaev@math.umu.se*
www.umu.se

Department of Forest Economics, Swedish University of Agricultural Sciences, Umeå, SE-90 183, Sweden
E-mail: yuri.belyaev@sekon.slu.se
E-mail: bengt.kristrom@sekon.slu.se
www.slu.se

Centre for Environmental Resource Economics (CERE), Umeå, SE-90 183, Sweden
E-mail: bengt.kristrom@sekon.slu.se
www.cere.se

A new approach, to analysis of statistical data collected in contingent valuation surveys, is introduced. In data collecting, randomly sampled respondents may freely self-select any interval that contains their willingness-to-pay (WTP-) point. Here, the presence of self-selected intervals, with rounded ends, is an essential feature. We avoid put restrictions on dependencies between the self-selected intervals and their associated WTP-points. Two-step sampling designs are proposed. Based on an estimated coverage probability, a practical rule to stop sampling in the first step is suggested. Then division intervals are generated and they are used on the second step of data collecting. A recursion is applied to maximize the log likelihood corresponding to data collected on the second step. The consistent maximum likelihood (ML-) estimates of WTP-distribution, projected on the division intervals, are found. Their accuracies may be consistently estimated by resampling methods.

Keywords: interval rounded data, estimable characteristics, maximisation likelihood, recursion, resampling.

Introduction

In surveys, it is useful to allow the respondent to respond to a certain type of questions with a freely chosen interval, rather than forcing him or her to state a point (or to choose one out of several presented brackets). One reason is that the respondent might not be completely certain about the quantity he or she is asked about. In such cases, the individual might refuse to answer the question, see Refs. 1, 2, 3 and 4 for evidence supporting this claim. Here we consider contingent valuation data, in which randomly selected respondents may freely self-select any interval of choice that contains their willingness-to-pay (WTP)-points. It is known that respondent finds it difficult to report WTP-points, and therefore self-selected intervals is a natural application to this case. The self-selected intervals can be considered as a kind of censoring of the true WTP-points. In the survival analysis it is typically assumed that the censoring intervals are independent of such points and cover only some of the points, see Refs. 5, 6, 7. In Ref. 8 existence of a common distribution of WTP-points in rescaled self-selected interval is assumed. In this paper we allow the stated intervals to depend on the unobserved positions of their WTP-points. We avoid putting restrictions on the type of dependence between self-selected intervals and their associated WTP-points. To make it possible we introduce a new two-step approach. We thus employ the first step to create an additional question about WTP which we use to fine-grain the information in the second step. The questions, stated to sampled respondents in the second step, are based on the data collected in the first step in a way that has not been explored before in the survey literature.

Our aim is to find consistent estimates related to the distribution of these unobservable WTP-points. We propose sampling designs and corresponding statistical models that allow a dependency between the self-selected WTP-intervals and the positions of their WTP-points.

Basic assumptions

Let $\mathfrak{P}$ be a population of individuals. We call randomly sampled from $\mathfrak{P}$ individuals as respondents. Our approach is based on three assumptions, of which we first discuss two (the third being statistical in nature).

Assumption 0.1. *Each respondent might not be aware of the exact location of the true WTP-point. The respondents may freely state intervals containing their true WTP-points. The ends of stated intervals may be rounded, e.g. to simple sums of coins or paper values of money.*

Assumption 0.2. *The true WTP-points are independent of question mode, i.e. the content of a question does not change the true WTP-point in the self-selected WTP-interval.*

Assumption 0.3. *The pairs of true WTP-points and the stated self-selected WTP-intervals, corresponding to different sampled individuals, are values of independent identically distributed (i.i.d.) random variables (r.v.s).*

The presence of rounded self-selected intervals, mentioned in Assumption 0.1, and anchoring are essential features of collected data.

Assumption 0.2 is important. From some perspectives, this assumption is very strong, given the evidence that exists on the differing results between types of valuation questions. But let us come back to it later and let us now turn to the statistical models.

Sampling design

We consider the following *two-step plan* of data collecting. On the *first step* randomly sampled individuals will be suggested to state self-selected intervals containing their true WTP-points. We assume that the fraction of sampled individuals is negligibly small.

Let n randomly sampled respondents have stated intervals $\mathbf{y}_1^n = \{\mathbf{y}_1, ..., \mathbf{y}_n\}$, $\mathbf{y}_i = (y_{Li}, y_{Ri}]$ containing their WTP-points. Due to rounding, the same intervals can be stated by different respondents. We consider n as a fixed non-random number.

From our assumptions it follows that we can consider stated intervals $\mathbf{y}_1^n = \{\mathbf{y}_1, ..., \mathbf{y}_n\}$ as a realization of a multinomial random process $\{\mathbf{Y}_i\}_{i\geq 1}$ with the discrete time parameter $i = 1, 2, ...$ The r.v.s $\{\mathbf{Y}_i\}_{i\geq 1}$ are i.i.d. and the set of their values is a set containing all possible to be stated self-selected intervals $\mathcal{U}_{all} = \{\mathbf{u}_\alpha : P[\mathbf{Y}_i = \mathbf{u}_\alpha] > 0, \alpha \in A\}$.

The set $\mathcal{U}_{all}$ can contain many different intervals u_α but due to rounding their number is finite. There is a discrete probability distribution $p_\alpha = P[\mathbf{Y}_i = \mathbf{u}_\alpha], \alpha \in A$. The set $\mathcal{U}_{all}$ and the probability distribution $\{p_\alpha, \alpha \in A\}$ are not known. In our context α is an integer index identifying $\mathbf{u}_\alpha$, i.e. $\mathbf{u}_{\alpha'} \neq \mathbf{u}_{\alpha''}$ if $\alpha' \neq \alpha''$.

All $m(n) \leq n$ different intervals in $\mathbf{y}_1^n$ can be ordered by the values of their ends. We write $\mathbf{y}_{i_1} < \mathbf{y}_{i_2}$ if either $y_{Li_1} < y_{Li_2}$, or $y_{Li_1} = y_{Li_2}$ but $y_{Ri_1} < y_{Ri_2}$. Then $\mathbf{y}_{i'_1} < \mathbf{y}_{i'_2} < ... < \mathbf{y}_{i'_{m(n)}}$ and we let $\mathbf{u}_{hn} = \mathbf{y}_{i'_h}$ for all $\mathbf{y}_i$ identical with $\mathbf{y}_{i'_h}$. The collected data can be written as a list:

$$\mathbf{d1}_n = \{..., \{h, \{u_{Lh,n}, u_{Rh,n}\}, t_{h,n}\}, ...\},$$

where ordering indexes $h = 1, ..., m(n)$ of different intervals $\mathbf{u}_{h,n} = (u_{Lh,n}, u_{Rh,n}]$ depend on the collected data $\mathbf{y}_1^n$.

Let $\mathcal{U}_{m(n),n} = \{\mathbf{u}_{1,n}, ..., \mathbf{u}_{m(n),n}\} \subseteq \mathcal{U}_{all}$ be the set of different stated WTP-intervals and $\mathbf{t}(n) = \{t_{1,n}, ..., t_{m(n),n}\}$ be the numbers $t_{h,n} = \sum_{i=1}^{n} I[\mathbf{y}_i = \mathbf{u}_{h,n}]$ of times that $\mathbf{u}_{h,n}$ was stated in the data $\mathbf{y}_1^n$.

We are interested in estimating the fraction of respondents, in the whole population $\mathfrak{P}$ of interest, who will state the WTP-intervals already collected in the set $\mathcal{U}_{m(n),n}$. Let $p_c(n)$ be probability of the event that the last WTP-interval, $\mathbf{y}_n = \mathbf{u}_{h_n,n}$, in $\mathbf{y}_1^n$, has been stated by at least one i-th of the previous respondents, $i < n$. Then $t_{h_n,n} \geq 2$. Let H_n be the r.v. to state an WTP-interval $\mathbf{u}_{H_n,n}$ by the n-th respondent. The expectation $E[I[t_{H_n,n} \geq 2]] = p_c(n)$. Therefore $I[t_{h_n,n} \geq 2]$ can be considered as a value of an unbiased estimate of $p_c(n)$, and $\mathbf{t}(n)$ is a sufficient statistic with components $t_{h,n}, h = 1, ..., m(n)$. Due to uniformly random sampling (simple random sampling without replacement) of respondents from $\mathfrak{P}$, $\mathbf{t}(n)$ is invariant with respect to any permutation of intervals in $\mathbf{y}_i^n$. Hence, after averaging $I[t_{H_n,n} \geq 2]$ given $\mathbf{t}(n)$, we obtain the enhanced unbiased estimate $\hat{p}_c(n) = r(n)/n$ of $p_c(n)$, where $r(n) = \sum_{h=1}^{m(n)} t_{h,n} I[t_{h,n} \geq 2]$.

We reduce our problem to a classical urn problem with $r(i)$ white and $n - r(i)$ black balls. Let $\mathcal{U}_{m(i),i}(1) = \{\mathbf{u}_h : \mathbf{u}_h \in \mathcal{U}_{m(i),i}, t_{hi} = 1\}, \mathcal{U}_{m(i),i}(2) = \{\mathbf{u}_h : \mathbf{u}_h \in \mathcal{U}_{m(i),i}, t_{hi} \geq 2\}, \mathcal{U}_{m(i),i} = \mathcal{U}_{m(i),i}(1) \cup \mathcal{U}_{m(i),i}(2)$.

If $\mathbf{y}_{i+1} \in \mathcal{U}_{m(i),i}(2)$ then $\Delta_c(i) = \hat{p}_c(i+1) - \hat{p}_c(i) = (i - r(i))/(i(i+1))$. If $\mathbf{y}_{i+1} \in \mathcal{U}_{m(i),i}(1)$ then $\Delta_c(i) = (2i - r(i))/(i(i+1))$. If $\mathbf{y}_{i+1} \notin \mathcal{U}_{m(i),i}$ then $\Delta_c(i) = -r(i)/(i(i+1))$.

By calculating $\hat{p}_c(i)$ for each $i \leq n$ we can observe the evolution of $\hat{p}_c(i)$, see Fig. 1. The probability $p_c(n)$ is increasing in n. Hence, we can use $\hat{p}_c(n)$ as a lower bound estimate for $p_c(n')$, for any $n' > n$. Then we may interpret $\hat{p}_c(n)100\%$ as a lower bound estimate of the percentage of all individuals in $\mathfrak{P}$ who being sampled would state a WTP-interval $\mathbf{u}_h \in \mathcal{U}_{m(n),n}$. We call $p_c(n)$ a *coverage probability.* The decision to stop data collection on the first step can thus depend on the value of $\hat{p}_c(n)$. If this value is not sufficiently close to 1, and if it is possible to prolong data collection, we have information about the value of so doing. If the collection of data has been stopped with $n_1 \geq n$ and $\hat{p}_c(n_1) = r(n_1)/n_1$ then inferences about the WTP-distribution will only correspond to the subset of individuals who will state WTP-intervals from $\mathcal{U}_{m(n_1),n_1}$.

Henceforth, the number n_1 of the randomly sampled, on the first step, respondents is fixed and we suppress indexes, e.g. we will write $\mathcal{U}_m = \{\mathbf{u}_1, ..., \mathbf{u}_m\}$ instead of $\mathcal{U}_{m(n_1),n_1} = \{\mathbf{u}_{1,n_1}, ..., \mathbf{u}_{m(n_1),n_1}\}$, and t_h instead of t_{h,n_1}.

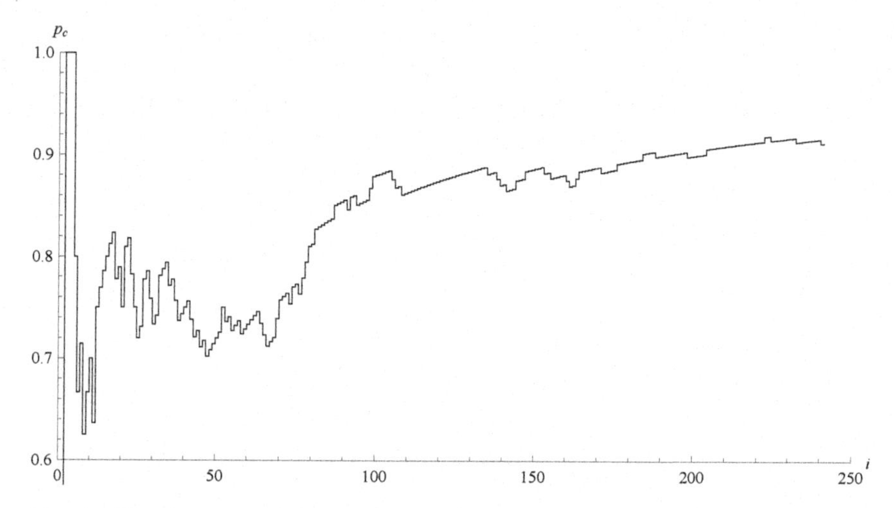

Fig. 1. The dynamics of the coverage probability estimates $\hat{p}_c(i)$ as a function of i sequentially selected respondents. The coverage probability estimates the fraction of respondents in the whole population who being selected would state self-selected WTP-intervals already stated by the first i selected respondents.

Each interval $\mathbf{u}_h \in \mathcal{U}_m$ can be written as the union of disjoint intervals $\mathbf{v}_1, ..., \mathbf{v}_k$, $\mathbf{v}_j = (v_{Lj}, v_{Rj}], v_{Rj} \leq v_{Lj+1}$. These intervals are ordered non-empty intersections of all intervals $\mathbf{u}_h \in \mathcal{U}_m$. We call them *division* intervals and we denote $\mathcal{V}_k = \{\mathbf{v}_1, ..., \mathbf{v}_k\}$.

Given our assumptions we consider respondents' WTP-points $\{x_i\}$ as values of i.i.d. r.v.s $\{X_i\}$. We need to estimate the d.f. $F[x] = P[X_i \leq x]$, or equivalently, we estimate the survival function (s.f.) $S[x] = 1 - F[x]$ of the true WTP-distribution. Let $H_i, i = 1, ..., n_1$, be i.i.d. r.v.s, $H_i = h$ given that, on the first step, the i-th respondent has stated interval $\mathbf{u}_h$ containing his/her WTP-point x_i. We let $w_h = P[\{H_i = h\} \cap \{X_i \in \mathbf{u}_h\}]$, $w_h \leq p_h = P[X_i \in \mathbf{u}_h], h = 1, ..., m$.

The events $\{H_i = h\} \cap \{X_i \in \mathbf{u}_h\}$ are observable. The events $\{X_i \in \mathbf{u}_h\}$ are, however, not observable. The probability of the number of times $t_1, ..., t_m$ to state $\mathbf{u}_1, ..., \mathbf{u}_m$ is $\prod_{h=1}^{m} w_h^{t_h}$, i.e. we have a multinomial distribution. The corresponding normed log likelihood (llik) is

$$\text{llik}[w_1, ..., w_m | t_1, ..., t_m] = \sum_{h=1}^{m} \frac{t_h}{n_1} \text{Log}[w_h], \quad \sum_{h=1}^{m} t_h = n_1.$$

The maximum of llik over $w_h \geq 0, \sum_{h=1}^{m} w_h = 1$, is attained at $\check{w}_h = t_h/n_1, h = 1, ..., m$. Note that $-\sum_{h=1}^{m} \check{w}_h \text{Log}[\hat{w}_h]$ is the empirical Entropy of the multinomial distribution with probabilities $\{\hat{w}_1, ..., \hat{w}_m\}$.

Let us now introduce the *second step* of data collection. We prolong random sampling of new (not yet sampled) individuals from the population $\mathfrak{P}$. In the second step each individual is to state an interval containing his/her WTP-point. If the stated interval does not belong to $\mathcal{U}_m$ then we do not include it in the collected data. If the stated self-selected interval belongs to $\mathcal{U}_m$ then this respondent is asked to select from the division $\mathcal{V}_k$ an interval $\mathbf{v}_j \in \mathcal{V}_k, \mathbf{v}_j \subseteq \mathbf{u}_h$, containing his/her true WTP-point. The respondent may well abstain from answering this second question and we will cater for this event in the simulations.

The collected data will be the list of triples $\mathbf{d2}_{n.2} = \{\mathbf{z}_1, ..., \mathbf{z}_{n.2}\}$, $\mathbf{z}_i = \{i, \mathbf{u}_{h_i}, NA\}$ or $\{i, \mathbf{u}_{h_i}, \mathbf{v}_{j_i}\}$, NA is "no answer" to the additional question. In short we call these triples *singles* and *pairs.* We suppose that $n._2$ is sufficiently large and that any $\mathbf{v}_j \in \mathcal{V}_k$ was stated many times. The size $n._2$ of the data collected in the second step should be significantly larger than n_1.

Statistical model and corresponding log likelihood

Now we consider r.v.s X_i and probabilities of events for individuals who being asked would state intervals included in $\mathcal{U}_m$.

Let us consider the conditional probabilities

$$w_{hj} = P[H_i = h \mid X_i \in \mathbf{v}_j].$$

We call a pair of h, j *compatible* if $\mathbf{v}_j \subseteq \mathbf{u}_h$. It is convenient to define the following two subsets of intervals' indexes

$$\mathcal{C}_h = \{j : \mathbf{v}_j \subseteq \mathbf{u}_h\} \text{ and } \mathcal{D}_j = \{h : \mathbf{v}_j \subseteq \mathbf{u}_h\}, j = 1, ..., k, h = 1, .., m.$$

It is not possible to identify the true WTP-distribution if we only have the data $\mathbf{d1}_{n_1}$ obtained on the first step. For identification of the true WTP-distribution on the division intervals $\mathcal{V}_k$ it is necessary to consistently estimate the conditional probabilities $\mathbb{W} = (w_{hj})$ by using extended empirical data. We obtain this identification asymptotically as the size of data collected in the second step is growing without bounds. The values of probabilities w_{hj} for a given $j = 1, ..., k$ reflect behavior of individual's selecting $\mathbf{u}_h \supseteq \mathbf{v}_j$, if $X_i = x_i \in \mathbf{v}_j$. We have

$$w_h = \sum_{j \in \mathcal{C}_h} w_{hj} q_{trj},$$

where $q_{trj} = P[X_i \in \mathbf{v}_j]$, is the true probability of the event $\{\mathbf{X}_i \in x_j\}$.

There are many hypothetical behavior models of respondents, e.g.

BM1 is the behavior model of indifferent respondents with $w_{hj} = 1/d_j, h \in \mathcal{D}_j,\ j = 1, ..., k,\ d_j$ is the size of $\mathcal{D}_j$.

BM2 is the behavior model of respondents who with $\mathbf{v}_j$ containing their WTP-point select to state an interval $\mathbf{u}_h$ in which $\mathbf{v}_j$ is the last division interval.

BM3 is the behavior model of respondents who, with $\mathbf{v}_j$ containing their WTP-point, select to state $\mathbf{u}_h, h \in \mathcal{D}_j$, proportionally to frequencies $w_h, h = 1, ..., m$. Here $\breve{w}_h = t_h/n_1$ and

$$\hat{w}_{hj} = \frac{t_h}{\left(\sum_{h' \in \mathcal{D}_j} t_{h'}\right)}. \tag{1}$$

We have little information about the process of selecting intervals but within our approach we do not need to know the true behavioral model. The collected data with pairs $\{i, \mathbf{u}_{h_i}, \mathbf{v}_{j_i}\}$ provides a simple way to estimate $q_{trj} = P[X_i \in \mathbf{v}_j], j = 1, ..., k$. Let $c_{pj} = \sum_{i=1}^{n_{\cdot 2}} I[\mathbf{z}_i = \{i, \mathbf{u}_{h_i}, \mathbf{v}_j\}],\ c_{phj} = \sum_{i=1}^{n_{\cdot 2}} I[\mathbf{z}_i = \{i, \mathbf{u}_h, \mathbf{v}_j\}]$, $n_{s2} = \sum_{h=1}^{m} t_{sh}$, $n_{p2} = \sum_{j=1}^{k} c_{pj}$, $n_{\cdot 2} = n_{s2} + n_{p2}$. The low indexes s and p correspond to *singles* and *pairs*, $t_{sh} = \sum_{i=1}^{n_{\cdot 2}} I[\mathbf{z}_i = \{i, \mathbf{u}_h, NA\}]$.

We consider $n_{p2}/n_{\cdot 2}$ as a frequency in the Binomial process with stated singles and pairs and $n_{p2}/n_{\cdot 2} \to \alpha_p \in (0, 1)$ almost surely (a.s.) as $n_{\cdot 2} \to \infty$. We can also consider the conditional Binomial process with pairs containing $\mathbf{v}_j$ together with any $\mathbf{u}_h \supseteq \mathbf{v}_j, \mathbf{u}_h \in \mathcal{U}_m$. To avoid the variety subsets of division intervals with $q_{trj} = 0$ we consider the case with all $q_{trj} > 0, j = 1, ..., k$. The cases with some of $q_{trj} = 0$ may be considered similarly.

The strongly consistent estimates of q_{trj} are

$$\breve{q}_{pj} = \frac{c_{pj}}{n_{p2}} \to q_{trj}, \quad \text{a.s.} \quad j = 1, ..., k, \quad \text{as } n_{\cdot 2} \to \infty. \tag{2}$$

We can also use the pairs $\{\mathbf{u}_{h_i}, \mathbf{v}_{j_i}\}$ to obtain the following strongly consistent estimates of w_{hj}

$$\hat{W}_{mk} = \{\hat{w}_{hj} = \frac{c_{phj}}{c_{pj}}, \quad h \in \mathcal{D}_j, \quad j = 1, ..., k\}. \tag{3}$$

We can write the following estimate of the normed by $n_{\cdot 2}$ log likelihood function of $\mathbf{q}_k = \{q_1, ..., q_k\}$ corresponding to the all data, with pairs and singles, collected on the second step

$$\text{llik}[\mathbf{q}_k \mid \hat{W}_{mk}, \mathbf{d2}_{n.2}] = \frac{n_{s2}}{n_{.2}} \sum_{h=1}^{m} \frac{t_{sh}}{n_{s2}} \text{Log}\left[\sum_{j \in \mathcal{C}_h} \hat{w}_{hj} q_j\right]$$

$$+ \frac{n_{p2}}{n_{.2}} \sum_{j=1}^{k} \frac{c_{pj}}{n_{p2}} \text{Log}[q_j] + \frac{1}{n_{.2}} \sum_{i=1}^{n_2} \text{Log}[\hat{w}_{h_i j_i}] I[\mathbf{z}_i = \{i, \mathbf{u}_{h_i}, \mathbf{v}_{j_i}\}]. \quad (4)$$

Recall that t_{sh} is the number of stated times $\mathbf{u}_h$ in the part of data with singles $\{i, \mathbf{u}_{h_i}, NA\}$.

The set of all possible values $q_j, j = 1, ..., k$, is the $(k-1)$-dimensional polyhedron

$$S_{k-1} = \{\mathbf{q}_k : 0 \le q_j \le 1, \sum_{j=1}^{k} q_j = 1\}.$$

Let $\partial \mathcal{S}_{k-1}$ be all boundary points of $\mathcal{S}_{k-1}$. The log likelihood function (4) is concave on $\mathcal{S}_{k-1} \setminus \partial \mathcal{S}_{k-1}$.

We use the method with Lagrange multipliers, see Ref. 9, and obtain a variant of an algorithm in the form of recursive iterations.

We have obtained the following recursion, $r = 1, 2, ...,$

$$q_j^{(r+1)} = \frac{n_{p2}}{n_{s2} + n_{p2}} q_j^{(1)} + \frac{n_{s2}}{n_{s2} + n_{p2}} \sum_{h \in \mathcal{D}_j} \frac{\hat{w}_{hj} q_j^{(r)}}{\sum_{j' \in \mathcal{C}_h} \hat{w}_{hj'} q_{j'}^{(r)}} \hat{w}_{sh}, \quad (5)$$

$$q_j^{(1)} = \check{q}_{pj}, \quad \hat{w}_{sh} = \frac{1}{n_{s2}} \sum_{i=1}^{n_{.2}} I[z_i = \{i, \mathbf{u}_h, NA\}].$$

$q_j^{(1)} = q_j^{(1)}(n_{p2}), q_j^{(r)} = q_j^{(r)}(n_{p2}, n_{s2}), \hat{w}_{sh} = \hat{w}_{sh}(n_{s2}), j = 1, ..., k, h = 1, ..., m$. If n_{p2} and n_{s2} are sufficiently large then

$$\hat{w}_{sh} > 0, \quad \sum_{j' \in \mathcal{C}_h} \hat{w}_{hj'} q_j^{(1)} > 0 \quad \text{and} \quad q_j^{(1)} = \check{q}_{pj} > 0,$$

for each $h = 1, ..., m, j = 1, ..., k$. The recursion can be applied for obtaining the ML-estimators of probabilities $\mathbf{q}_{tr}$ maximizing llik (4).

From strong consistency $q_j^{(1)} \to q_{trj}$ a.s. $n_{.2} \to \infty, j = 1, ..., k$, and (5) it follows that $\mathbf{q}_{trk} \in \mathcal{K}_{\alpha_p} = \{\mathbf{q}_k : q_j \ge q_{trj}\alpha_p/2, \sum_{j=1}^{k} q_j = 1\} \subset \mathcal{S}_{k-1} \setminus \partial \mathcal{S}_{k-1}$. For all sufficiently large $n_{.2}$ all $\mathbf{q}_k^{(r)} \in \mathcal{K}_{\alpha_p}, r \ge 1$. Due to concavity of (4) the sequence $\{\mathbf{q}_k^{(r)}\}_{r \ge 1}$ has only one limit stationary point of the recursion (5) in the compact set $\mathcal{K}_{\alpha_p}$ where (4) has its maximum. We summarize these results in the following theorem.

Theorem 0.1. *Suppose that Assumptions 0.1–0.3 are valid and the sizes* n_{p2} *and* n_{s2} *of collected pairs and singles are growing unboundedly as* $n_{\cdot 2} \to \infty$. *Then for any sufficiently large* n_{p2} *and* n_{s2} *the ML-estimator* $\check{\mathbf{q}}_k = \{\check{q}_1, ..., \check{q}_k\}$ *(based on the data* $\mathbf{d2}_{n_{\cdot 2}}$ *with singles and pairs) of the log likelihood (4) exists, is strongly consistent, and can be found as the stationary point of recursion (5).*

Hence, we can find consistent ML-estimates $\check{q}_j, j = 1, ..., k$, of the projected WTP-distribution on the division intervals $\mathbf{v}_j \in \mathcal{V}_k$, i.e. $\check{q}_j \to q_{trj}$ a.s. $j = 1, ..., k$, if n_{p2} and n_{s2} are growing. Note that this inference can be only applied to the respondents in $\mathfrak{P}$ who will state $\mathbf{u}_h \in \mathcal{U}_m$.

Mean WTP, i.e. $m_{tr1} = E[X_i]$, is one of the most essential parameters in cost-benefit analysis, see Ref. 3. Here, however, rounding of WTP-intervals excludes the possibility to find an unbiased estimate of m_{tr1}. Therefore, we introduce an approximation m_{1tr} of the following consistently estimable *medium mean* of the WTP-distribution

$$mm_{1tr} = v_{L1} + \sum_{j=1}^{k} \left(q_{tr,j+1}^{k} + \frac{1}{2} q_{trj} \right) (v_{Rj} - v_{Lj}),$$

$$q_{tr,j+1}^{k} = \sum_{i=j+1}^{k} q_{tri}, \quad q_{tr,k+1} = 0.$$

We have the following strongly consistent ML-estimate of mm_{1tr}

$$\check{mm}_{1n_{\cdot 2}} = v_{L1} + \sum_{j=1}^{k} \left(\check{q}_{j+1}^{k} + \frac{1}{2} \check{q}_j \right) (v_{Rj} - v_{Lj}).$$

$\check{q}_j$ is the ML-estimator of q_{trj} with the likelihood given in (4), $\check{q}_{j+1}^{k} = \sum_{i=j+1}^{k} \check{q}_i$, $j = 1, ..., k$, $\check{q}_{k+1} = 0$. Note that $\check{mm}_{1n_{\cdot 2}}$ is the area below the broken line, with points $\{v_{Rj}, \check{q}_{j+1}^{k}\}, j = 1, ..., k-1$ and $\{v_{Rk}, 0\}$ on the estimated WTP-survival function, connected by intervals.

Numerical experiment

We consider the data collected and used in the cost-benefit analysis[2,3] as if they were realized on the fist step of data collection. All self-selected intervals are given in the set $\mathcal{U}_m$, and all division intervals in $\mathcal{V}_k, m = 46, k = 23$. The estimate of the coverage probability is $\hat{p}_c(n_1), \simeq 91\%, n_1 = 241$. Its evolution is in Fig. 1. In the second step we use the subsets $\mathcal{C}_h = \{j : \mathbf{v}_j \subseteq \mathbf{u}_h\}$ and $D_j = \{h : \mathbf{v}_j \subseteq \mathbf{u}_h\}$. In order to illustrate how our approach is working we have used simulated data as if they were obtained on the second step of data collection.

For a simulation of the data collected on the second step it is necessary to know the distribution of true WTP-points $\{X_i\}$ and the behavior model describing decisions of respondents to state self-selected WTP-intervals. We have supposed that the behavior model is BM3 and the WTP-distribution is a p_{WE}-mixture WE of the Weibull $W(a, b)$ and the Exponential $E(m_1)$ distributions with parameters $p_{WE} = 0.8160, a = 74.8992, b = 1.8374, m1 = 254.7344$. The WTP-survival function is

$$sf_{WTP}[x] = p_{WE}e^{-(x/a)^b} + (1 - p_{WE})e^{-x/m_1}. \tag{6}$$

To illustrate our approach we have simulated $n_{\cdot 2} = 9000$ triples as if they were collected on the second step. Here, we used $\hat{w}_{hj}$ defined in (1), and the WE-mixture (6). The numbers of singles and pairs were $n_{s2} = 5856$ and $n_{p2} = 3144$. We find the corresponding estimates $\check{q}_{pi}, \hat{w}_{hj}$, given in (2) and (3) and $\hat{w}_{sh} = t_{sh}/n_{sh}$. It is also possible to use estimate $\tilde{w}_{sh} = n_{\cdot 2}^{-1} \sum_{i=1}^{n_2} I[z_i = \{i, h, \_\}]$, "_" means NA or any $\mathbf{v}_j$. Now, we apply recursion (5) to approximate by $q_j^{(r)}$ the ML-estimators $\check{q}_j$ corresponding the log likelihood (4). It was sufficient to use 5 iteration of the recursion. Observe that we in this numerical example know the survival function (6). In Fig. 2 the s.f. is shown together with pairs of true points $\{v_{Rj}, sf_{WTP}[v_{Rj}]\}$

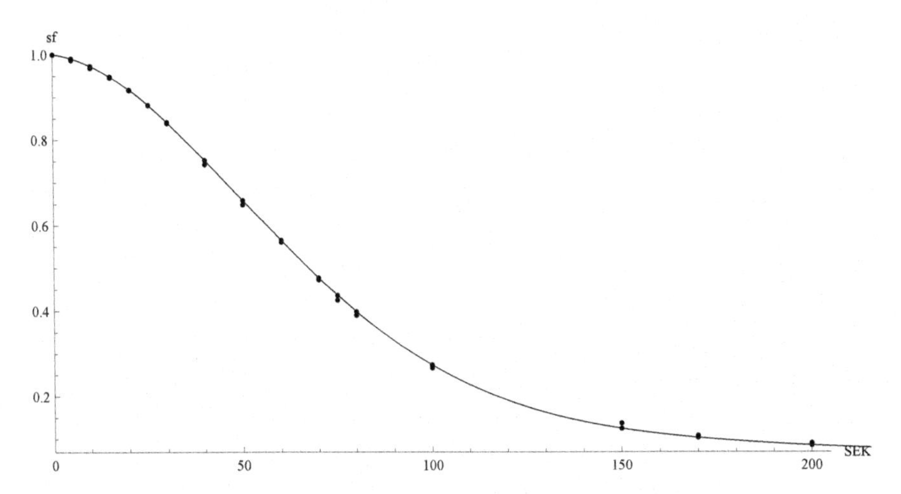

Fig. 2. Part of the "true" WTP-survival function (6) line together with points on the line and with the ML-estimated values of the s.f. at the ends of division intervals. The list of all $n_{\cdot 2} = 9000$ triples and 5 times of recursion iterations were used.

and $\{v_{Rj}, \sum_{j'=j+1}^{k} q_{j'}^{(5)}\}, j = 1, ..., k$, and $\{0, 1\}$, $\{v_{Rk}, 0\}$. The points obtained after 5 iterations are rather close to the corresponding true points on the true s.f., and their accuracy is much better than if we would used (2) in the points $\{v_{Rj}, \sum_{j'=j+1}^{k} \check{q}_{pj}\}, j = 1, ..., k$.

The considered statistical model with the log likelihood (4) satisfies all regularity conditions, see Refs. 10–13. Hence, we can use resampling methods in estimation accuracy of ML-estimates and functions of them, see Refs. 14–17. For example, we can evaluate accuracy of the medium mean value estimate $\check{mm}_{1n.2} = 104.22$ which is the linear function of the ML-estimates $\check{q}_{jn.2}$ approximated by $q_j^{(5)}, j = 1, ..., k$. In our example, for all respondents who will state self-selected intervals from $\mathcal{U}_m$, the true WTP-mean value $m_{1tr} = 101.17$ and the true medium mean value $mm_{1tr} = 103.85$. Bias is 2.58. In Fig. 3 we present QQ-plot with two distribution of deviations: the true distribution of deviations $\check{mm}_{1n.2} - mm_{1tr}$ (dashed line), and its estimate obtained by resamplings of the list with all triples (solid line).

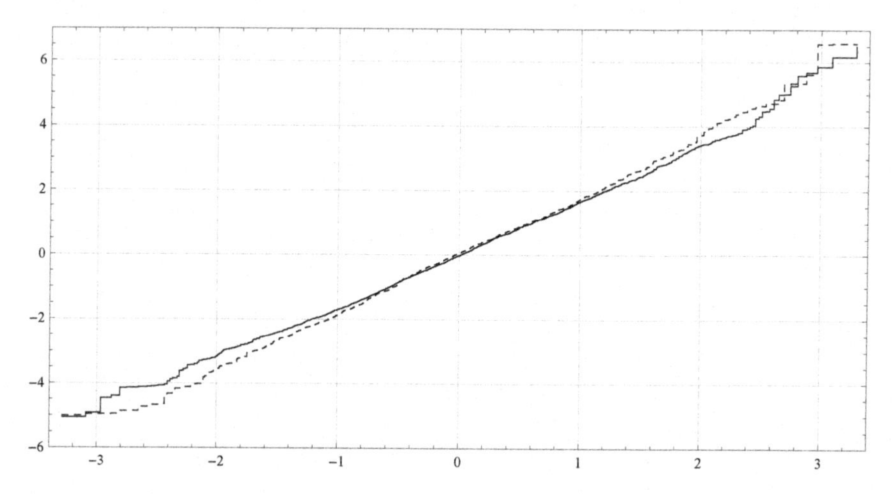

Fig. 3. Two distributions, of the deviations of medium mean WTP-values from their average values. The dashed line corresponds to case with $C = 2000$ copies of true deviations $\check{mm}_1^{(5)c} - \frac{1}{C}\sum_{c'=1}^{C} \check{mm}_1^{(5)c'}$. The solid line corresponds to case with $C = 2000$ copies of deviations $\check{mm}_1^{\star(5)c} - \frac{1}{C}\sum_{c'=1}^{C} \check{mm}_1^{\star(5)c'}$ obtained via resamplings.

Conclusion

This paper introduces several new ideas of potential use in survey research. We have proposed a natural extension of the self-selected intervals approach to information elicitation in surveys, see Ref. 1 for a summary of our previous research on self-selected intervals. The self-selected intervals are very general and we believe they can be a natural candidate in many types of surveys. For additional motivation, examples and different methods, see Ref. 4. The key novelty proposed in this paper is the introduction of a second step of data collection. We thus solve the problem of consistent estimation of the WTP-distribution by adding more information in a second step, where this step uses the division intervals.

We also propose a stopping rule for sampling of respondents, on the first step, which is very practical. The investigator is to suggest a coverage probability, say 0.95, which when reached means that there is a 5% chance that on the second step any next sampled self-selected interval will be different from the ones already stated on the first step.

Theorem 1 states that the proposed non-parametric ML-estimates of $q_{trj}, j = 1, ..., k$, are consistent. The accuracy of these estimators can be consistently estimated by resampling, see Fig. 3.

In future research, we will take the next natural step, which is to use the two-step approach for real data. In addition, there is a need to introduce explanatory variables.

Acknowledgements

The research presented in this paper was carried out as a part of two research projects. It was initiated by the R&D program "Hydropower - Environmental impacts, mitigation measures and costs in regulated waters". It has been established and financed by Elforsk, the Swedish Energy Agency, the National Board of Fisheries and the Swedish Environmental Protection Agency, `www.vattenkraftmiljo.nu`. The second is PlusMinus, sponsored by the Swedish Environmental Protection agency, `plusminus.slu.se`. Our thanks to the Department of Mathematics and Mathematical Statistics, Umeå University, to Dr. M. Ekström for useful comments, and to referees for their remarks.

References

1. Y. Belyaev and B. Kriström, *Approach to Analysis of Self-selected Interval Data,* CERE Working Paper, 2010:2, (Umeå, 2010). http://www.cere.se/

2. C. Håkansson, Cost-Benefit Analysis and Valuation Uncertainty, Doctoral Thesis, Swedish University of Agricultural Sciences, (Umeå, Sweden, 2007).
3. C. Håkansson, A new valuation question - Analysis of and insights from interval open ended data in contingent valuation, *Environmental and Resource Economics,* **39**, 2, (2008), pp. 175–188.
4. C. F. Manski and F. Molinari, Rounding Probabilistic Expectations in Surveys, *Journal of Business and Economic Statistics,* **28**, (2010), pp. 219-231.
5. S. R. Jammalamadaka and V. Mangalam, Non-parametric estimation for middle-censored data, *J. Nonparametr. Statist.* **15**, (2003), pp. 253–265.
6. J. P. Klein and M. L. Moeschberger, *Survival Analysis: Techniques for Censored and Truncated Data,* (Springer-Verlag, N. Y. 1997).
7. B. M. Turnbull, The empirical distribution function with arbitrarily grouped, censored and truncated data, *J. Roy. Statist. Soc.* Ser. B 38, (1996), pp. 290–295.
8. M. Ekström, *Nonparametric estimation for classic and interval open-ended data in contingent valuation,* Research report, 2010:7, Swedich University of Agricultural Sciences, (Umeå, Sweden, 2010). http://biostochastics.slu.se
9. J. Brinkhuis and V. Tihomirov *Optimization: Insights and Applications,* (Princeton University Press, Princeton and Oxford, 2005).
10. Y. Belyaev and L. Nilsson, *Parametric Maximum Likelihood Estimators and Resampling,* Research Report, 15, Department of Mathematical Statistics, Umeå University, (Umeå, Sweden, 1997).
11. T. S. Ferguson, *A Course in Large Sample Theory,* (Chapman & Hall, London, 1996).
12. E. L. Lehmann and G. Casella, *Theory of point estimation,* 2nd. ed., (Springer-Verlag, N. Y., 1998).
13. L. Nilsson, Parametric Estimation Using Computer Intensive Methods, PhD thesis, Umeå University, (Umeå, Sweden, 1998).
14. Y. Belyaev, *Necessary and sufficient conditions for consistency of resampling,* Research report, Centre of Biostochastics. Swedish University of Agricultutal Sciences, (Umeå, Sweden, 2003). http://biostochastics.slu.se
15. Y. Belyaev, *Resampling for Lifetime Distributions,* in *Encyclopedia of Statistics in Quality and Reliability,* Ed. F. Rupper, R. Kennet, F. W. Faltin, (John Wiley, Chichester, 2007), pp. 1653–1657.
16. Y. Belyaev and S. Sjöstedt-de Luna, Weakly approaching sequences of random distributions, *J. Applied Probability,* **37**, (2000), pp. 807–822.
17. A. C. Davison and D. Hinkley, *Bootstrap Methods and their Applications,* (Cambridge University Press, Cambridge, 1997).

OPTIMAL CLASSIFICATION OF MULTIVARIATE GRF OBSERVATIONS

K. DUČINSKAS* and L. DREIŽIENĖ**

Department of Statistics, Klaipėda University,
Klaipėda, LT-92294, Lithuania
**E-mail: kestutis.ducinskas@ku.lt*
***E-mail: l.dreiziene@gmail.com*
Tel.: +370 46 398850; fax: +370 46 398802
www.ku.lt

The problem of classifying a multivariate Gaussian random field (GRF) single observation into one of two populations specified by different parametric mean models and common intrinsic covariogram is considered. This paper concerns with classification procedures associated with the linear Bayes Discriminant Function (BDF) under the deterministic spatial sampling design. In the case of parametric uncertainty, the ML estimators of unknown parameters are plugged in the BDF. The actual risk and the approximation of the expected risk associated with aforementioned plug-in BDF (PBDF) are derived. This is the extension of the previous one to the case of general loss function and for complete parametric uncertainty, i.e. when mean and covariance functions parameters are unknown. The values of the derived approximation are examined for various combinations of parameters for the bivariate, stationary geometric anisotropic Gaussian random field with exponential covariance function sampled on regular 2-dimensional lattice.

Keywords: Bayes discriminant function; Covariogram; Gaussian random field; Actual risk; Training labels configuration.

1. Introduction

The paper deals with evaluation of the performance of spatial classification procedures based on Bayes discriminant functions. Saltyte-Benth and Ducinskas[1] derived the actual error rate and the asymptotic approximation of the expected error rate when classifying the observation of a multivariate Gaussian random field into one of two classes. The classes are specified by different regression mean models and common covariance function. The same approach to discrimination for feature observations having elliptically contoured distributions is implemented in Ref. 2. However, in both publications the observations to be classified are assumed to be independent from training samples. This is an unrealistic assumption particularly when the locations of observations to be classified are close to ones of training sample.

The first extension of previous results to the case when spatial correlations between Gaussian observations to be classified and observations in the training sample are not assumed equal to zero is done by Ducinskas[3]. Here only the trend parameters and scale parameter of the covariance function are assumed unknown. The case of multivariate GRF with known spatial correlation function is considered in Ref. 4. The extension to the complete parametric uncertainty was done in Ref. 5. The similar approximation of error rate is presented in Ref. 6.

In the present paper the approximation in the multivariate case of complete parametric uncertainty (all means and covariance function parameters are unknown) is implemented. We focus on the maximum likelihood (ML) estimators, since the inverse of the information matrix, associated with likelihood function of the training sample approximates the covariance matrix of these estimators well. The asymptotic properties of ML estimators showed by Mardia and Marshall[7] under increasing domain asymptotic framework and subject to some regularity conditions are essentially exploited. The asymptotic results sometimes yield useful approximations to finite-sample properties. The derived approximate expected risk (AER) is proposed as natural performance measure for PBDF.

By using the proposed AER, the performance of the PBDF is numerically analyzed in the case of stationary Gaussian random field on 2-dimensional regular lattice with geometrically isotropic exponential correlation function. Constant means, feature covariance matrix and anisotropy ratio are assumed unknown. The dependence of the values of derived AER on the statistical parameters such as the range parameter and the anisotropy ratio is investigated.

The results of numerical analysis give us the strong arguments for suggestion to use the derived formulas of risks as a performance criterion for the spatial classification procedures based on Bayes discriminant functions. The proposed BDF and PBDF could also be considered as the extension of widely used Bayesian methods to the restoration of image corrupted by spatial Gaussian noise (see Ref. 8, ch. 7.4).

2. The main concepts and definitions

The main objective of this paper is to classify a single observation of a multivariate GRF $\{Z(s) : s \in \mathbf{D} \subset \Re^2\}$.

The model of observation $Z(s)$ in population Ω_l is

$$Z(s) = B_l' x(s) + \omega(s), \; l = 1, \; 2 \tag{1}$$

where $x(s)$ is a $q \times 1$ vector of non random regressors and B_l is the $q \times p$ matrix of parameters.

The error term is generated by p-variate zero-mean stationary GRF $\{\omega(s) : s \in \mathbf{D}\}$ with intrinsic covariogram defined by the following model for all $s, u \in \mathbf{D}$

$$cov\{\omega(s), \omega(u)\} = \rho(s-u; \theta)\Sigma, \tag{2}$$

where $\rho(s - u; \theta)$ is the spatial correlation function and Σ is variance–covariance matrix with elements $\{\sigma_{ij}\}$ and $\theta \in \Theta$ is a parameter vector, Θ being an open subset of $\Re^r$.

In the casc when covariance function is known, the model Eqs. (1), (2) is called the universal kriging model (see Ref. 8, ch. 3).

For the given training sample, consider the problem of classification of the $Z_0 = Z(s_0)$ into one of two populations when

$$x'(s_0)B_1 \neq x'(s_0)B_2, s_0 \in \mathbf{D}.$$

Denote by $S_n = \{s_i \in \mathbf{D}; i = 1, ..., n\}$ the set of locations where the training sample $T' = (Z(s_1), ..., Z(s_n))$ is taken, and call it the set of training locations (STL). It specifies the spatial sampling design or spatial framework for the training sample.

We shall assume the deterministic spatial sampling design and all analyses are carried out conditionally on S_n.

Assume that the each training sample realization $T = t$ and S_n are arranged in the following way. The first n_1 components are observations of $Z(s)$ from Ω_1 and remaining $n_2 = n - n_1$ components are the observations of $Z(s)$ from Ω_2. So S_n is partitioned into union of two disjoint subsets, i. e. $S_n = S^{(1)} \cup S^{(2)}$, where $S^{(j)}$ is the subset of S_n that contains n_j locations of feature observations from Ω_j, $j = 1, \; 2$. So each partition $\xi(S_n) = \{S^{(1)}, S^{(2)}\}$ with marked labels determines training labels configuration (TLC).

For TLC $\xi(S_n)$, define the variable $d = |D^{(1)} - D^{(2)}|$, where $D^{(j)}$ is the sum of distances between the location s_0 and locations in $S^{(j)}$, $j = 1, \; 2$.

Then $n \times 2q$ design matrix of training sample T denoted by X is specified $X = X_1 \oplus X_2$, where symbol $\oplus$ denotes a direct sum of matrices and X_j is the $n_j \times q$ matrix of regressors for observations from Ω_j, $j = 1, 2$.

In what follows, we assume that STL S_n and TLC ξ are fixed. This is the case, when spatial classified training data are collected at fixed locations (stations).

Then the model of T is

$$T = XB + W, \tag{3}$$

where X is the $n \times 2q$ design matrix, $B' = (B_1', B_2')$ is the $p \times 2q$ matrix of means parameters and W is the $n \times p$ matrix of random errors that has matrix-variate normal distribution i.e.

$$W \sim N_{n \times p}(0, P \otimes \Sigma).$$

Here P denotes the spatial correlation matrix for T.

Denote by ρ_0 the vector of spatial correlations between Z_0 and observations in T and set

$$\alpha_0(\theta) = P^{-1}\rho_0, \ k(\theta) = 1 - \rho_0'\alpha_0.$$

Notice that in population Ω_l, the conditional distribution of Z_0 given $T = t$ is Gaussian, i.e.

$$(Z_0|T = t, \Omega_l) \sim N_p(\mu_{lt}^0, \Sigma_{0t}), \tag{4}$$

where conditional means μ_{lt}^0 are

$$\mu_{lt}^0(B, \theta) = E(Z_0|T = t; \Omega_l) = B_l'x_0 + (t - XB)'\alpha_0, \ l = 1, \ 2 \tag{5}$$

and conditional covariance matrix Σ_{0t} is

$$\Sigma_{0t}(\theta, \eta) = V(Z_0|T = t; \Omega_l) = k(\theta)\Sigma. \tag{6}$$

The marginal squared Mahalanobis distance between populations for observation taken at location $s = s_0$ is

$$\Delta^2 = (\mu_1^0 - \mu_2^0)'\Sigma^{-1}(\mu_1^0 - \mu_2^0),$$

where $\mu_l^0 = B_l'x_0$, $l = 1, 2$.

This is the Mahalanobis distance between marginal distributions of observation Z_0 modeled by Eqs. (1), (2). It can be used for determination of the level of statistical separation between populations.

The squared Mahalanobis distance between conditional distributions of Z_0 for given $T = t$ is specified by

$$\Delta_n^2 = (\mu_{1t}^0 - \mu_{2t}^0)'\Sigma_{0t}^{-1}(\mu_{1t}^0 - \mu_{2t}^0).$$

Then using Eqs. (5), (6) yields

$$\Delta_n^2 = \Delta^2/k(\theta).$$

It should be noted that Δ_n does not depend on the realizations of T and depends only on STL S_n.

Let $L(i,j)$ be the loss incurred by making decision j if the true class is i.

Under the assumption of complete parametric certainty of populations and for known finite nonnegative losses $\{L(i,j),\ i,j = 1,2\}$, the BDF minimizing the risk of the classification is formed by log ratio of conditional likelihoods.

Then for $T = t$, BDF is the linear function with respect to Z_0 (see Ref. 9)

$$W_t(Z_0) = \Big(Z_0 - (\mu_{1t}^0 + \mu_{2t}^0)/2\Big)' \Sigma_{0t}^{-1}(\mu_{1t}^0 - \mu_{2t}^0) + \gamma, \tag{7}$$

where $\gamma = \ln(\pi_1^*/\pi_2^*)$.

Here $\pi_j^* = \pi_j\Big(L(j,3-j) - L(j,j)\Big)$, $j = 1,\ 2$, where π_1, π_2 $(\pi_1 + \pi_2 = 1)$ are prior probabilities of the populations Ω_1 and Ω_2, respectively.

Lemma 2.1. *The risk for the BDF $W_t(Z_0)$ is*

$$R_0 = \sum_{j=1}^{2}\Big(\pi_j^*\Phi\big(-\Delta_n/2 + (-1)^j\gamma/\Delta_n\big) + \pi_j L(j,j)\Big), \tag{8}$$

where $\Phi(\cdot)$ is the standard normal distribution function.

Proof of lemma. Recall that under the Definition 1 presented in Ref. 5

$$R_0 = \sum_{i=1}^{2}\sum_{j=1}^{2}\pi_i L(i,j) P_{it}\big((-1)^j W_t(Z_0) < 0\big),$$

where the probability measure P_{it} is based on the conditional distribution of Z_0 given $T = t$, Ω_i specified in Eqs. (4)-(6).

In the population Ω_j, the conditional distribution of $W_t(Z_0)$ given $T = t$ is normal with mean

$$E_j(W_t(Z_0)) = (-1)^{j+1}\Delta_n^2/2 + \gamma$$

and variance

$$Var_j(W_t(Z_0)) = \Delta_n^2,\ j = 1,\ 2.$$

Then using the properties of multivariate normal distribution we complete the proof of lemma. □

In what follows, the risk R_0 will be called Bayes risk. Note that it does not depend on realizations of T.

In practical applications not all statistical parameters of populations are known. Then the estimators of unknown parameters can be found from the training sample. The estimators of unknown parameters are plugged into BDF, and PBDF is used for the classification. In this paper we assume that true values of parameters B, Σ and θ are unknown (complete parametric uncertainty).

Let $H = (I_q, I_q)$ and $G = (I_q, -I_q)$, where I_q denotes the identity matrix of order q. Replacing parameters by their estimators in Eq. (7) and using Eqs. (5), (6), we get the PBDF

$$\hat{W}_T(Z_0) = (Z_0 - (T - X\hat{B})'\hat{\alpha}_0 - \hat{B}'H'x_0/2)'\hat{\Sigma}^{-1}\hat{B}'G'x_0/\hat{k} + \gamma. \quad (9)$$

Definition 2.1. Given $T = t$, the actual risk for PBDF $\hat{W}_T(Z_0)$ is defined as

$$\hat{R}(t) = \sum_{i=1}^{2}\sum_{j=1}^{2} \pi_i L(i,j) P_{it}\big((-1)^j \hat{W}_t(Z_0) < 0\big).$$

Lemma 2.2. *The actual risk for $\hat{W}_t(Z_0)$ specified in Eq. (9) is*

$$\hat{R}(t) = \sum_{j=1}^{2} \big(\pi_j^* \Phi(\hat{Q}_j) + \pi_j L(j,j)\big), \quad (10)$$

where

$$\hat{Q}_l = (-1)^l \frac{(x_0'(B_l - H\hat{B}/2) + \alpha_0'(t - XB) - \hat{\alpha}_0'(t - X\hat{B}))\hat{\Sigma}^{-1}\hat{B}'G'x_0 + \gamma\hat{k}}{\sqrt{x_0'G\hat{B}\hat{\Sigma}^{-1}\Sigma\hat{\Sigma}^{-1}\hat{B}'G'x_0 k}}. \quad (11)$$

Proof of lemma. It is obvious that in the population Ω_l the conditional distribution of PBDF $\hat{W}_T(Z_0)$ given $T = t$ is Gaussian, i.e.,

$$\hat{W}_T(Z_0)|T = t,\ \Omega_l \sim N(m_{tl}, v_{tl}), \quad (12)$$

where

$$m_{tl} = \big(x_0'(B_l - H\hat{B}_t/2) + \alpha_0' X(\Delta\hat{B}_t)\big)\hat{\Sigma}_t^{-1}\hat{B}_t'G'x_0/k + \gamma \quad (13)$$

$$\nu_{tl} = x_0' G \hat{B}_t \hat{\Sigma}_t^{-1} \Sigma \hat{\Sigma}_t^{-1} \hat{B}_t' G' x_0 / k. \tag{14}$$

The proof of lemma is completed and formulas (10), (11) are obtained by using the equations (9), (12)-(14) and the definition 2.1. □

Definition 2.2. The expectation of the actual risk with respect to the distribution of T is called the expected risk (ER) and is designated as $E_T(\hat{R}(T))$.

More comprehensive information about the actual and expected risks for classification into arbitrary number of populations one can find in Ref. 10.

The ER is useful in providing a guide to the performance of plug-in classification rule before it is actually formed from the training sample. The ER is the performance measure to the PBDF, similarly to the mean squared prediction error (MSPE) is the performance measure to the plug-in kriging predictor (see Ref. 11). The approximations of MSPE for plug-in kriging were suggested in previous papers (e.g. Ref. 12). These approximations are used for the spatial sampling design criterion for prediction (see Refs. 13, 14). These facts strengthen the motivation for the derivation of closed form expressions for the AER associated with PBDF.

3. The asymptotic approximation of ER

We will use the maximum likelihood (ML) estimators of parameters based on the training sample. The asymptotic properties of ML estimators established by Mardia and Marshall[7] under increasing domain asymptotic framework and subject to some regularity conditions are essentially exploited. Henceforth, denote by MM the regularity conditions of Theorem 1 from paper above.

Set $\beta = vec(B)$, $\eta = vech(\Sigma)$, $P_\theta = \partial vec P / \partial \theta'$, $\dim \beta = q_0 = 2qn$, $\dim \eta = m = p(p+1)/2$.

The log-likelihood function of T (specified in Eq. (3)) is

$$\Lambda(\beta, \eta, \theta) = const - \big(p \ln |P| + n \ln |\Sigma| + tr\big(P^{-1}(T - XB)\Sigma^{-1}(T - XB)\big)\big)/2.$$

Then the information matrices for the corresponding parameters are

$$J_\beta = (X'P^{-1}X) \otimes \Sigma^{-1}, \; J_\eta = nD_p'(\Sigma^{-1} \otimes \Sigma^{-1})D_p/2,$$

$$J_\theta = pP_\theta'(P^{-1} \otimes P^{-1})P_\theta/2,$$

where D_p is the duplication matrix of order $p^2 \times (p(p+1)/2)$.

Note that $J_{\eta\theta} = E_T(\partial^2\Lambda(\Psi)/\partial\eta\partial\theta')$ and the above information matrices are evaluated at the true values of parameters β, η and θ.

It is easy to obtain that

$$J_{\eta\theta} = \big(D'_m(\Sigma^{-1} \otimes \Sigma^{-1})vec(\Sigma)\big) \otimes \big(vec'P(P^{-1} \otimes P^{-1})P_\theta/2\big).$$

Denote by $J = \begin{pmatrix} J_\eta & J_{\eta\theta} \\ J_{\theta\eta} & J_\theta \end{pmatrix}$ and $V = J^{-1} = \begin{pmatrix} V_\eta & V_{\eta\theta} \\ V_{\theta\eta} & V_\theta \end{pmatrix}$ the information matrix and inverse of information matrix, respectively. Under some regularity condition, the matrix V is an approximate covariance of the ML estimators of covariance function parameters.

Using properties of the multivariate Gaussian distribution it is easy to prove that

$$\hat{\beta} \sim AN_{q_0}(\beta, V_B), \hat{\eta} \sim AN_r(\theta, V_\eta), \hat{\theta} \sim AN_m(\theta, V_\theta), \tag{15}$$

where symbols $AN(.,.)$ mean approximate normal distribution.

Let $R_\beta^{(k)}$, $R_\eta^{(k)}, R_\theta^{(k)}, k = 1,\ 2$ denote the k-th order derivatives of $\hat{R}(T)$ with respect to $\hat{\beta}$ and $\hat{\theta}$ evaluated at the point $\hat{\beta} = \beta,\ \hat{\eta} = \eta,\ \hat{\theta} = \theta$ and let $R_{\beta\eta}^{(2)}, R_{\beta\theta}^{(2)}, R_{\eta\theta}^{(2)}$ denote the matrices of the second order derivatives of $\hat{R}(T)$ with respect to $\hat{\beta},\ \hat{\eta}$ and $\hat{\theta}$ evaluated at the point $\hat{\beta} = \beta,\ \hat{\eta} = \eta,\ \hat{\theta} = \theta$.

The following assumption is made:

(A1) The training sample T and estimators $\hat{\eta},\ \hat{\theta}$ are statistically independent.

The restrictive assumption (A1) is exploited intensively by many authors (see Refs. 13, 14), since Abt[12] showed that finer approximations of MSPE considering the correlation between T and $\hat{\theta}$ do not give better results.

Let $A_\theta = \partial\hat{\alpha}_0/\partial\hat{\theta}'$ be the $n \times k$ matrix of partial derivatives evaluated at the point $\hat{\theta} = \theta$ and let $\varphi(\cdot)$ be the standard normal distribution density function.

Set $\Lambda = X'\alpha_0 - (H'/2 + \gamma G'/\Delta_n^2)x_0,\ R_0 = (X'P^{-1}X)^{-1}$,
$\varphi_1 = \varphi(-\Delta_n/2 - \gamma/\Delta_n),\ \Delta\mu = \mu_1^0 - \mu_2^0$.

Theorem 3.1. *Suppose that observation Z_0 is to be classified by BPDF and let conditions (MM) and assumption (A1) hold. Then the asymptotic approximation of ER is*

$$\begin{aligned} AER = R_0 + \pi_1^*\varphi_1\big(\Lambda' R_0 \Lambda \Delta_n/k + (p-1)x_0' G R_0 G' x_0/(k\Delta_n) \\ + tr(F_1 V_\eta) + tr(F_2 V_\theta) + 2tr(F_3 V_{\eta\theta})\big)/2 \end{aligned} \tag{16}$$

where R_0 is the Bayes risk specified in Eq. (8),

$$\begin{aligned} F_1 = D_p'\big((\Sigma^{-1}\Delta\mu\Delta\mu'\Sigma^{-1} \otimes \Sigma^{-1}\Delta\mu\Delta\mu'\Sigma^{-1})\gamma^2 k/\Delta^4 \\ + \Sigma^{-1}\Delta\mu\Delta\mu'\Sigma^{-1} \otimes (\Sigma^{-1} - \Sigma^{-1}\Delta\mu\Delta\mu'\Sigma^{-1}/\Delta^2)\big)D_p/(\Delta\sqrt{k}) \end{aligned} \tag{17}$$

$$F_2 = (tr(A_\theta' P A_\theta V_\theta)\Delta^2 + \gamma^2 k_\theta' V_\theta k_\theta)/(\Delta^3\sqrt{k}) \tag{18}$$

$$F_3 = D_p'(\Sigma^{-1}\Delta\mu \otimes \Sigma^{-1}\Delta\mu)(\gamma^2 k_\theta)/(\Delta^4\sqrt{k}) \tag{19}$$

Proof. Expanding $\hat{R}(T)$ in the Taylor series around point $\hat{\beta} = \beta, \hat{\eta} = \eta$ and $\hat{\theta} = \theta$ up to the second order and taking expectation with respect to the approximate distribution specified in Eq. (15) we have

$$\begin{aligned} E_T(\hat{R}(T)) = R_0 + E_T\big((\Delta\hat{\beta})' R_\beta^{(2)} \Delta\hat{\beta} + 2(\Delta\hat{\theta})' R_{\theta\eta}^{(2)} \Delta\hat{\eta} \\ + (\Delta\hat{\theta}') R_\theta^{(2)} (\Delta\theta) + (\Delta\hat{\eta})' R_\eta^{(2)} (\Delta\hat{\eta})\big)/2 + R_3, \end{aligned} \tag{20}$$

where $\Delta\hat{\beta} = \hat{\beta} - \beta$, $\Delta\hat{\eta} = \hat{\eta} - \eta$, $\Delta\hat{\theta} = \hat{\theta} - \theta$ and R_3 is the reminder term. After doing matrix algebra we have

$$\begin{aligned} R_\beta^{(2)} = \pi_1^* \varphi_1\big((\Sigma^{-1}\Delta\mu\Delta\mu'\Sigma^{-1} \otimes \Lambda\Lambda')/k \\ + (\Sigma^{-1} - \Sigma^{-1}\Delta\mu\Delta\mu'\Sigma^{-1}/\Delta^2) \otimes (G' x_0 x_0' G)\big)/\Delta\sqrt{k} \end{aligned} \tag{21}$$

and

$$\begin{aligned} R_\eta^{(2)} = \pi_1^* \varphi_1 D_p'((\Sigma^{-1}\Delta\mu\Delta\mu'\Sigma^{-1} \otimes \Sigma^{-1}\Delta\mu\Delta\mu'\Sigma^{-1})\gamma^2 k/\Delta^4 \\ + \Sigma^{-1}\Delta\mu\Delta\mu'\Sigma^{-1} \otimes (\Sigma^{-1} - \Sigma^{-1}\Delta\mu\Delta\mu'\Sigma^{-1}/\Delta^2))D_p/(\Delta\sqrt{k}) \end{aligned} \tag{22}$$

$$E(R_\theta^{(2)}) = \pi_1^* \varphi_1 (A_\theta' P A_\theta \Delta^2 + \gamma^2 k_\theta k_\theta')/(\Delta^3\sqrt{k}) \tag{23}$$

$$E(R_{\theta\eta}^{(2)}) = \pi_1^* \varphi_1 (D_m'(\Sigma^{-1}\Delta\mu \otimes \Sigma^{-1}\Delta\mu)(\gamma^2 k_\theta)/(\Delta^4\sqrt{k}) \tag{24}$$

Then by using the assumption (A1) and Eqs. (23), (24) and replacing $E_T(\Delta\hat{\theta}\Delta\hat{\theta}')$ and $E_T(\Delta\hat{\eta}\Delta\hat{\theta}')$ by their approximations V_θ and $V_{\theta\eta}$ we get the following approximations

$$\begin{aligned} E((\Delta\hat{\theta})' R_\theta^{(2)} (\Delta\hat{\theta})) \\ \cong \pi_1^* \varphi_1 (tr(A_\theta' P A_\theta V_\theta)\Delta_0^2 + \gamma^2 k_\theta' V_\theta k_\theta)/(\Delta^3\sqrt{k}), \end{aligned} \tag{25}$$

$$\begin{aligned} E((\Delta\hat{\eta})' R_{\theta\eta}^{(2)} (\Delta\hat{\theta})) \\ \cong \pi_1^* \varphi_1 \gamma^2 (tr(D_m'(\Sigma^{-1}\Delta\mu \otimes \Sigma^{-1}\Delta\mu) k_\theta V_{\theta\eta}))/(\Delta^4\sqrt{k}). \end{aligned} \tag{26}$$

Then using Eqs. (21), (22), (25), (26) in the right-hand side of Eq. (20), and dropping the reminder term, we complete the proof of Theorem 3.1. □

4. Example and discussions

A numerical example is considered to investigate the influence of the statistical parameters of populations to the proposed AER in the finite (even small) training sample case.

With an insignificant loss of generality the cases with $n_1 = n_2, \pi_1 = \pi_2 = 0,5$ and $L(i,j) = 1 - \delta_{ij},\ i,j = 1,\ 2$, are considered.

In this example, observations are assumed to arise from stationary Gaussian random field with the constant mean and intrinsic covariance function given by $C(h) = \rho(h)\Sigma$. We consider the case of exponential geometric anisotropic correlation function $\rho(h)$ specified by

$$\rho(h) = \exp\Big\{ - \sqrt{h_x^2 + \lambda^2 h_y^2}\Big/\alpha\Big\}.$$

Here λ denotes the unknown anisotropy ratio and α denotes known range parameter (see Ref. 11). The anisotropy angle is assumed to be fixed $\phi = \pi/2$.

So, we consider the case with $\theta = \lambda$.

Set $s_i - s_j = h(ij) = (h_x(ij), h_y(ij)),\ i,j = 0,1,..,n$.

Then $P_\theta = -vec(P \circ D)$, where $\circ$ denotes Hadamard product of matrices. The elements of $n \times n$ matrix $D = (d_{ij})$ are specified by

$$d_{ii} = 0,\ d_{ij} = \lambda h_y^2(ij)\Big/\Big(\alpha\sqrt{h_x^2(ij) + \lambda^2 h_y^2(ij)}\Big),\ i \neq j;\ i,j = 1,..,n.$$

Also notice that

$$A_\theta = -P^{-1}(P_\theta \alpha_0 + \rho_0 \circ E),$$

where $E' = (d_{01}, ..., d_{0n})$ and $k_\theta = \rho_0' \circ E' - \rho_0' A_\theta$.

Assume that $\mathbf{D}$ is regular 2-dimensional lattice with unit spacing. Consider the case $s_0 = (1,1)$ and fixed STL S_8 contains 8 second-order neighbors of s_0.

Consider TLC ξ_1 for the training sample specified by

$$\xi_1 = \{S^{(1)} = \{(1,2),(2,2),(2,1),(2,0)\},\ S^{(2)} = \{(1,0),(0,0),(0,1),(0,2)\}\}.$$

This is symmetric TLC since $d = 0$.

For this TLC the values of AER specified in Eqs. (16)-(19) are calculated for various values of parameters $\lambda,\ \alpha$. The results of calculations with $\Delta = 1$ are presented in Table 1.

Analyzing the contents of Table 1 we can conclude that for TLC ξ_1 the AER decreases when the range parameter α increases. The character of the dependence between AER and anisotropy ratio λ depends on the value of α. Those conclusions will be significant for the solution of applied problems.

Table 1. Values of AER for TLC ξ_1 with $\Delta = 1$ and various α and λ.

				α			
λ	0.8	1.2	1.6	2.0	2.4	2.8	3.2
1	0.30726	0.27095	0.24097	0.21575	0.19421	0.17555	0.15922
2	0.31400	0.28144	0.25502	0.23240	0.21267	0.19525	0.17974
3	0.31362	0.28081	0.25490	0.23292	0.21372	0.19672	0.18153
4	0.31314	0.28047	0.25472	0.23291	0.21387	0.19699	0.18190
5	0.31255	0.28020	0.25460	0.23288	0.21390	0.19708	0.18202
6	0.31200	0.27994	0.25450	0.23285	0.21391	0.19711	0.18207
7	0.31154	0.27972	0.25440	0.23282	0.21391	0.19712	0.18210
8	0.31116	0.27953	0.25431	0.23278	0.21390	0.19714	0.18212
9	0.31084	0.27938	0.25424	0.23275	0.21389	0.19714	0.18212
10	0.31058	0.27925	0.25417	0.23272	0.21388	0.19713	0.18213

Hence the results of numerical analysis give us strong arguments to expect that proposed approximation of the expected risk (expected error rate) could be effectively used for the performance evaluation of classification procedures and for the optimal designing of the spatial training samples.

References

1. J. Saltyte-Benth and K. Ducinskas, Linear discriminant analysis of multivariate spatial-temporal regressions, *Scandinavian Journal of Statistics* **32**, 281–294 (2005).
2. A. Batsidis and K. Zografos, Discrimination of observations into one of two elliptic populations based on monotone training samples, *Metrika* **64** (2006).
3. K. Ducinskas, Approximation of the expected error rate in classification of the Gaussian random field observations, *Statistics and Probability Letters* **79**, 138–144 (2009).
4. K. Ducinskas, Error rates in classification of multivariate Gaussian random field observation, *Lithuanian Mathematical Journal* **51** 4, 477–485 (2011).
5. K. Ducinskas and L. Dreiziene, Supervised classification of the scalar Gaussian random field observations under a deterministic spatial sampling design, *Austrian Journal of Statistics* **40** 1, 2, 25–36 (2011).
6. K. Ducinskas and L. Dreiziene, Application of Bayes Discriminant Functions to Classification of the Spatial Multivariate Gaussian Data, *Procedia Environmental Sciences* **7**, 212–217 (2011).
7. K. V. Mardia and R. J. Marshall, Maximum likelihood estimation of models for residual covariance in spatiall regression, *Biometrika* **71**, 135–146 (1984).
8. N. A. C. Cressie, *Statistics for spatial data* (Wiley, New York, 1993).
9. G. J. McLachlan, *Discriminant analysis and statistical pattern recognition* (Wiley, New York, 2004).
10. K. Ducinskas, An asymptotic analysis of the regret risk in discriminant analysis under various training schemes, *Lithuanian Mathematical Journal* **37** 4, 337–351 (1997).

11. P. J. Diggle, P. J. Ribeiro and O. F. Christensen, An introduction to model-based geostatistics, *Lecture notes in statistics* **173**, 43–86 (2002).
12. M. Abt, Estimating the prediction mean squared error in Gaussian stochastic processes with correlation structure, *Scandinavian Journal of Statistics* **26**, 563–578 (1999).
13. D. L. Zimmerman, Optimal network design for spatial prediction, covariance parameter estimation, and empirical prediction, *Environmetrics* **17**, 635–652 (2006).
14. Z. Zhu and M. L. Stein, Spatial sampling design for prediction with estimated parameters, *Journal of Agricultural, Biological, and Environmental Statistics* **11** 1, 24–44 (2006).

MULTIVARIATE EXPONENTIAL DISPERSION MODELS

B. JØRGENSEN

Department of Mathematics and Computer Science,
University of Southern Denmark, Campusvej 55, 5230 Odense M, Denmark
E-mail: bentj@stat.sdu.dk

J. R. MARTÍNEZ

FAMAF, Universidad Nacional de Córdoba,
Ciudad Universitaria, 5000 Córdoba, Argentina
E-mail: jmartine@famaf.unc.edu.ar

We develop a new class of multivariate exponential dispersion models with a fully flexible correlation structure, and present multivariate versions of the Poisson, binomial, negative binomial, gamma, inverse Gaussian, and Tweedie distributions, some of which extend existing bivariate or multivariate models of these types. The new models are constructed using an extended convolution method, which interpolates between the set of fully correlated pairs of variables and the set of independent marginals of the prescribed form. The models have an additive as well as a reproductive form. The additive form is particularly suitable for discrete data, whereas the reproductive form is more suitable for continuous data, having properties similar to the multivariate normal distribution. Multivariate exponential dispersion models are useful as error distributions for multivariate generalized linear models for the purpose of modelling multivariate non-normal data.

Keywords: Convolution method; Multivariate binomial distribution; Multivariate dispersion model; Multivariate gamma distribution; Multivariate generalized linear model; Multivariate inverse Gaussian distribution; Multivariate negative binomial distribution; Multivariate Poisson distribution; Multivariate Tweedie distribution.

1. Introduction

The aim of this investigation is to construct multivariate distributions with margins of prescribed form, having a fully flexible correlation structure. Our motivation is to be able to model multivariate non-normal data by some form of multivariate generalized linear model, which in turn requires a suitable form of multivariate exponential dispersion model. Such multivariate models were introduced by Jørgensen,[1] who considered the general problem of constructing multivariate dispersion models, extending the univariate case studied in detail by Jørgensen.[2]

There is no shortage of multivariate non-normal distributions available, but the class of multivariate exponential dispersion models proposed by Jørgensen[1] has the advantage of possessing a fully flexible correlation structure, while having margins of the prescribed form. A multivariate exponential dispersion model in its reproductive form (cf. Sec. 5) is parametrized by a k-vector of means $\boldsymbol{\mu}$ and a symmetric positive-definite $k \times k$ dispersion matrix $\boldsymbol{\Sigma}$, thereby resembling the multivariate normal distribution in much the same way that univariate exponential dispersion models resemble the univariate normal distribution.[2]

Multivariate exponential dispersion models of additive form (cf. Sec. 3) are constructed by an extended convolution method, which interpolates between the set of fully correlated pairs of variables and the set of independent margins of the prescribed form. This method explores the convolution property of conventional additive exponential dispersion models (cf. Sec. 2) in order to generate the desired number of parameters, namely k means and $k(k+1)/2$ variance and covariance parameters. Multivariate additive exponential dispersion models are particularly suitable for discrete data, and include multivariate versions of the Poisson, binomial and negative binomial distributions (cf. Sec. 4).

The reproductive form of multivariate exponential dispersion model is constructed by applying the so-called duality transformation to a given multivariate additive exponential dispersion model. The reproductive form is particularly suited for continuous data, and includes the multivariate normal distribution as a special case, along with new multivariate forms of gamma, inverse Gaussian and other Tweedie distributions (cf. Secs. 5–6).

2. Multivariate dispersion models

In order to set the stage for the construction of multivariate exponential dispersion models, we begin by a short review of multivariate proper dispersion models and ordinary exponential dispersion models.

2.1. *Multivariate proper dispersion models*

Following Jørgensen[1] and Jørgensen and Lauritzen,[3] we define a multivariate proper dispersion model $\mathrm{PD}_k(\boldsymbol{\mu}, \boldsymbol{\Sigma})$ as having probability density function of the form

$$f(\boldsymbol{y};\boldsymbol{\mu},\boldsymbol{\Sigma}) = a(\boldsymbol{\Sigma})b(\boldsymbol{y})\exp\left[-\frac{1}{2}r^\top(\boldsymbol{y};\boldsymbol{\mu})\boldsymbol{\Sigma}^{-1}r(\boldsymbol{y};\boldsymbol{\mu})\right] \text{ for } \boldsymbol{y}\in\mathbb{R}^k, \quad (1)$$

where $r(\boldsymbol{y};\boldsymbol{\mu})$ is the k-vector of deviance residuals (signed square-root deviances), $\boldsymbol{\mu}$ denotes the $k\times 1$ position vector, $\boldsymbol{\Sigma}$ is the $k\times k$ dispersion matrix, and a and b are suitable functions of $\boldsymbol{\Sigma}$ and $\boldsymbol{y}$, respectively. The multivariate normal distribution is a special case of (1) obtained for $\boldsymbol{r}(\boldsymbol{y};\boldsymbol{\mu}) = \boldsymbol{y}-\boldsymbol{\mu}$ and $b(\boldsymbol{y}) = 1$. A number of examples of (1) were considered by Jørgensen and Lauritzen[3] and Jørgensen.[1]

The multivariate proper dispersion model density (1) imitates the multivariate normal density in a fairly direct way via the quadratic form of deviance residuals appearing in the exponent of (1). The construction of multivariate proper dispersion models relies on identifying cases where the vector of deviance residuals $r(\boldsymbol{y};\boldsymbol{\mu})$ is such that the normalizing constant $a(\boldsymbol{\Sigma})$ depends on the parameters only through $\boldsymbol{\Sigma}$. However, this technique does not work for exponential dispersion models, calling for an alternative method of construction.

2.2. *Ordinary exponential dispersion models*

We shall now review some requisite details of exponential dispersion models in the multivariate case, following Jørgensen.[4] The additive form of the k-variate exponential dispersion model is denoted $\boldsymbol{X}\sim \mathrm{ED}^*(\boldsymbol{\mu},\lambda)$, where $\boldsymbol{\mu}$ is the $k\times 1$ *rate vector* and λ is the positive *weight parameter* (sometimes called the index or convolution parameter). This model is defined by a probability density/mass function on $\mathbb{R}^k$ of the form

$$f^*(\boldsymbol{x};\boldsymbol{\theta},\lambda) = a^*(\boldsymbol{x};\lambda)\exp\left[\boldsymbol{x}^\top\boldsymbol{\theta} - \lambda\kappa(\boldsymbol{\theta})\right], \quad (2)$$

where the cumulant function $\kappa(\boldsymbol{\theta})$ depends on the canonical parameter $\boldsymbol{\theta}\in\Theta\subseteq\mathbb{R}^k$ related to $\boldsymbol{\mu}$ by (4) below, and $a^*(\boldsymbol{x};\lambda)$ is a suitable function of $\boldsymbol{x}$ and λ. In most cases we shall assume infinite divisibility in the following, which assures that the weight parameter λ has domain $\mathbb{R}_+$. We note that the model $\mathrm{ED}^*(\boldsymbol{\mu},\lambda)$ has $k+1$ parameters, of which the single parameter λ controls the covariance structure of the distribution, see below. Our main goal is, so to speak, to replace λ by a weight matrix $\boldsymbol{\Lambda}$ in order to obtain a fully flexible covariance structure with $k(k+1)/2$ variance and covariance parameters.

In much of the following we work with the cumulant generating function (CGF), and in particular the CGF for $\mathrm{ED}^*(\boldsymbol{\mu},\lambda)$ is given by

$$\lambda\kappa(\boldsymbol{s};\boldsymbol{\theta}) = \lambda\left[\kappa(\boldsymbol{s}+\boldsymbol{\theta}) - \kappa(\boldsymbol{\theta})\right] \text{ for } s\in\Theta-\boldsymbol{\theta}. \quad (3)$$

From now on it is understood that a given CGF is finite for the argument $\boldsymbol{s}$ belonging to a neighbourhood of zero, and hence characterizes the distribution in question. We also repeatedly use the fact that the mean vector and covariance matrix may be obtained as the first and second derivatives of the CGF at zero, respectively. In particular the mean vector and covariance matrix for $\boldsymbol{X}$ are $\lambda\boldsymbol{\mu}$ and $\lambda\boldsymbol{V}(\boldsymbol{\mu})$, respectively, where

$$\boldsymbol{\mu} = \dot{\kappa}(\boldsymbol{\theta}) \qquad \text{and} \qquad \boldsymbol{V}(\boldsymbol{\mu}) = \ddot{\kappa} \circ \dot{\kappa}^{-1}(\boldsymbol{\mu}). \tag{4}$$

Here dots denote derivatives, and the matrix $\boldsymbol{V}(\boldsymbol{\mu})$ is the so-called *unit variance function*. It also follows from (3) that the model $\mathrm{ED}^*(\boldsymbol{\mu}, \lambda)$ satisfies the following additive property:

$$\mathrm{ED}^*(\boldsymbol{\mu}, \lambda_1) + \mathrm{ED}^*(\boldsymbol{\mu}, \lambda_2) = \mathrm{ED}^*(\boldsymbol{\mu}, \lambda_1 + \lambda_2). \tag{5}$$

We note in passing that the additive property implies the existence of a k-variate Lévy process, i.e. a stochastic process with stationary and independent increments [2, pp. 83–84], such that $\boldsymbol{\mu}$ is the vector of rates for the process; hence the terminology rate vector for $\boldsymbol{\mu}$.

To each additive exponential dispersion model $\mathrm{ED}^*(\boldsymbol{\mu}, \lambda)$ the corresponding reproductive exponential dispersion model $\mathrm{ED}(\boldsymbol{\mu}, \sigma^2)$ is defined by the *duality transformation* $\boldsymbol{Y} = \lambda^{-1}\boldsymbol{X}$, or

$$\mathrm{ED}(\boldsymbol{\mu}, \lambda^{-1}) = \lambda^{-1}\mathrm{ED}^*(\boldsymbol{\mu}, \lambda), \tag{6}$$

say, where $\sigma^2 = \lambda^{-1}$ is called the *dispersion parameter*. From now on we refer to $\mathrm{ED}(\boldsymbol{\mu}, \sigma^2)$ and $\mathrm{ED}^*(\boldsymbol{\mu}, \lambda)$ as *ordinary* (as opposed to multivariate) exponential dispersion models. The model $\mathrm{ED}(\boldsymbol{\mu}, \sigma^2)$ satisfies the following reproductive property: if $\boldsymbol{Y}_1, \ldots, \boldsymbol{Y}_n$ are i.i.d. $\mathrm{ED}(\boldsymbol{\mu}, \sigma^2)$, then

$$\frac{1}{n}\sum_{i=1}^{n} \boldsymbol{Y}_i \sim \mathrm{ED}(\boldsymbol{\mu}, \sigma^2/n). \tag{7}$$

A summary of the additive and reproductive forms of exponential dispersion model is shown in Table 1 at the end of Sec. 3, including the multivariate forms $\mathrm{ED}_k^*(\boldsymbol{\mu}, \boldsymbol{\Lambda})$ and $\mathrm{ED}_k(\boldsymbol{\mu}, \boldsymbol{\Sigma})$ to be defined in Secs. 3 and 5 below.

3. Convolution method, additive form

3.1. *The bivariate case*

We shall now introduce the extended convolution method in the bivariate additive case. Suppose that we are given an ordinary bivariate additive $\mathrm{ED}^*(\boldsymbol{\mu}, \lambda)$ with CGF

$$(s_1, s_2)^\top \mapsto \lambda\kappa(s_1, s_2; \theta_1, \theta_2) = \lambda\left[\kappa(s_1 + \theta_1, s_2 + \theta_2) - \kappa(\theta_1, \theta_2)\right]. \tag{8}$$

From this model we define a multivariate additive exponential dispersion (bivariate case) by means of the following stochastic representation for the random vector $\boldsymbol{X}$:

$$\begin{bmatrix} X_1 \\ X_2 \end{bmatrix} = \begin{bmatrix} U_{11} \\ U_{12} \end{bmatrix} + \begin{bmatrix} U_1 \\ 0 \end{bmatrix} + \begin{bmatrix} 0 \\ U_2 \end{bmatrix}, \tag{9}$$

where the three terms on the right-hand side of (9) are assumed independent. More precisely, we define the joint CGF for $\boldsymbol{X}$ to be

$$K(s_1, s_2; \boldsymbol{\theta}, \boldsymbol{\Lambda}) = \lambda_{12}\kappa(s_1, s_2; \theta_1, \theta_2) + \lambda_1\kappa(s_1, 0; \theta_1, \theta_2) + \lambda_2\kappa(0, s_2; \theta_1, \theta_2), \tag{10}$$

which by (8) may also be expressed as follows:

$$\begin{aligned} &\lambda_{12}\kappa(s_1 + \theta_1, s_2 + \theta_2) + \lambda_1\kappa(s_1 + \theta_1, \theta_2) + \lambda_2\kappa(\theta_1, s_2 + \theta_2) \\ &- (\lambda_{12} + \lambda_1 + \lambda_2)\,\kappa(\theta_1, \theta_2). \end{aligned} \tag{11}$$

Here $\boldsymbol{\theta} = (\theta_1, \theta_2)^\top$, and $\boldsymbol{\Lambda}$ is the *weight matrix* defined by

$$\boldsymbol{\Lambda} = \begin{bmatrix} \lambda_{11} & \lambda_{12} \\ \lambda_{12} & \lambda_{22} \end{bmatrix},$$

with $\lambda_{ii} = \lambda_{12} + \lambda_i$ for $i = 1, 2$.

We interpret (10) as interpolating between independence ($\lambda_{12} = 0$) and the maximally correlated case ($\lambda_1 = \lambda_2 = 0$). We note that the margins have the same form as for $\lambda\kappa(s_1, s_2; \theta_1, \theta_2)$, as seen from the two marginal CGFs:

$$\begin{aligned} K(s_1, 0; \boldsymbol{\theta}, \boldsymbol{\Lambda}) &= (\lambda_{12} + \lambda_1)\,\kappa(s_1, 0; \theta_1, \theta_2) = \lambda_{11}\kappa(s_1, 0; \theta_1, \theta_2) \\ K(0, s_2; \boldsymbol{\theta}, \boldsymbol{\Lambda}) &= (\lambda_{12} + \lambda_2)\,\kappa(0, s_2; \theta_1, \theta_2) = \lambda_{22}\kappa(0, s_2; \theta_1, \theta_2). \end{aligned}$$

The construction hence preserves the form of the univariate margins, while replacing the single parameter λ by the three parameters of $\boldsymbol{\Lambda}$, for a total of five parameters.

Using the notation (4), we find that the mean vector for $\boldsymbol{X}$ is

$$\mathrm{E}(\boldsymbol{X}) = \begin{bmatrix} \lambda_{11}\dot{\kappa}_1(\theta_1, \theta_2) \\ \lambda_{22}\dot{\kappa}_2(\theta_1, \theta_2) \end{bmatrix} = \boldsymbol{\lambda\mu}, \tag{12}$$

say, where $\dot{\kappa}_i$ denotes the ith component of the first derivative of κ for $i = 1, 2$, and where $\boldsymbol{\lambda} = \operatorname{diag}\{\lambda_{11}, \lambda_{22}\}$ is a 2×2 diagonal matrix. We use the notation $\boldsymbol{X} \sim \mathrm{ED}_2^*(\boldsymbol{\mu}, \boldsymbol{\Lambda})$ for the multivariate additive exponential dispersion defined by (10), parametrized by the rate vector $\boldsymbol{\mu}$ and weight matrix $\boldsymbol{\Lambda}$.

The above construction is a special case of the variables-in-common method [5, Ch. 7], and provides a slight, but important extension of the ordinary convolution method, the latter being the special case where $U_{11} = U_{12}$ in (9). In fact the ordinary convolution method suffers from the problem that the equation $U_{11} = U_{12}$ implies $\mu_1 = \mu_2$, thereby in effect leaving us with only one rate parameter instead of the two (in general k) parameters that we require.

The covariance matrix for $\boldsymbol{X}$ takes the form

$$\mathrm{Var}(\boldsymbol{X}) = \begin{bmatrix} \lambda_{11}V_{11}(\boldsymbol{\mu}) & \lambda_{12}V_{12}(\boldsymbol{\mu}) \\ \lambda_{12}V_{21}(\boldsymbol{\mu}) & \lambda_{22}V_{22}(\boldsymbol{\mu}) \end{bmatrix} = \boldsymbol{\Lambda} \odot \boldsymbol{V}(\boldsymbol{\mu}) \tag{13}$$

where $\odot$ denotes the Hadamard (elementwise) product, and where the V_{ij} are elements of the unit variance function defined by (4). The correlation between X_1 and X_2 is

$$\mathrm{Corr}(X_1, X_2) = \frac{\lambda_{12}}{\sqrt{\lambda_{11}\lambda_{22}}} \frac{V_{12}(\boldsymbol{\mu})}{\sqrt{V_{11}(\boldsymbol{\mu})V_{22}(\boldsymbol{\mu})}},$$

which for given $\boldsymbol{\mu}$ varies between zero and $V_{12}(\boldsymbol{\mu})/\sqrt{V_{11}(\boldsymbol{\mu})V_{22}(\boldsymbol{\mu})}$, which may be either positive or negative, depending on the sign of $V_{12}(\boldsymbol{\mu})$. Ideally we would like the bound to be either $+1$ or -1, respectively, which is the case if

$$V_{12}(\boldsymbol{\mu}) = \pm\sqrt{V_{11}(\boldsymbol{\mu})V_{22}(\boldsymbol{\mu})}. \tag{14}$$

But if (14) is satisfied, then the matrix $\boldsymbol{V}(\boldsymbol{\mu})$ is singular, which in turn requires that the two variables U_{11} and U_{12} be linearly related, which is the case for the ordinary convolution method ($U_{11} = U_{12}$). In Secs. 4–6 we consider several examples that go beyond the ordinary convolution method in various ways.

At this point we should emphasize that the new model with CGF (10) is not in general a multivariate natural exponential family for $\boldsymbol{\Lambda}$ known, because K is not of the form $f(\boldsymbol{s} + \boldsymbol{\theta}) - f(\boldsymbol{\theta})$, as evident from the expression (11). We do of course start the construction from a natural exponential family, in the sense that (2) is a natural exponential family when λ is known. However, the apparent flexibility and generality of our method may be contrasted with the rather restricted options available for constructing multivariate natural exponential families, see e.g. Bar-Lev et al.[6] and references therein. In this sense, we have sacrificed the simplicity of the natural exponential family in order to obtain a fully flexible correlation structure.

3.2. *The trivariate and multivariate cases*

Before moving on to the full multivariate case it is convenient to discuss some important issues in connection with the trivariate case. Following

Jørgensen,[1] we define the trivariate random vector $\boldsymbol{X} = (X_1, X_2, X_3)^\top$ as the sum of six independent terms,

$$\begin{bmatrix} X_1 \\ X_2 \\ X_3 \end{bmatrix} = \begin{bmatrix} U_{11} \\ U_{12} \\ 0 \end{bmatrix} + \begin{bmatrix} U_{21} \\ 0 \\ U_{23} \end{bmatrix} + \begin{bmatrix} 0 \\ U_{32} \\ U_{33} \end{bmatrix} + \begin{bmatrix} U_1 \\ 0 \\ 0 \end{bmatrix} + \begin{bmatrix} 0 \\ U_2 \\ 0 \end{bmatrix} + \begin{bmatrix} 0 \\ 0 \\ U_3 \end{bmatrix}, \quad (15)$$

of which three terms are bivariate and three are univariate. However, rather than starting from a trivariate CGF $\kappa(s_1, s_2, s_3; \boldsymbol{\theta})$ as was done by Jørgensen,[1] it is natural to start from the three bivariate distributions corresponding to the first three terms of (15). For this construction to work, it is crucial that the univariate margins of the three bivariate terms are consistent, so that for example U_{11} and U_{21} have distributions that belong to the same class. In order to avoid the intricacies of such a construction in the general case, we concentrate here on the homogeneous case, where all three margins belong to the same class, for example a multivariate gamma distribution with gamma margins.

As illustrated by (12), each univariate margin depends, as a rule, on both canonical parameters θ_i and θ_j, say, which is counterproductive in terms of obtaining the above-mentioned consistency of the univariate margins. This prompts us to move to the parametrization based on the rate parameter μ. We shall hence impose the condition that all three margins must belong to a given univariate additive exponential dispersion model $\mathrm{ED}^*(\mu, \lambda)$ with CGF of the form $\lambda\kappa(s; \mu)$, say. This implies that the bivariate distributions with CGF $\kappa(s_1, s_2; \mu_1, \mu_2)$, say, must have margins satisfying $\kappa(s_1, 0; \mu_1, \mu_2) = \kappa(s_1; \mu_1)$ and $\kappa(0, s_2; \mu_1, \mu_2) = \kappa(s_2; \mu_2)$. We now define the trivariate exponential dispersion model $\mathrm{ED}_3^*(\boldsymbol{\mu}, \boldsymbol{\Lambda})$ via the joint CGF for the vector $\boldsymbol{X}$ as follows:

$$\begin{aligned} &K(s_1, s_2, s_3; \boldsymbol{\mu}, \boldsymbol{\Lambda}) \\ &\quad = \lambda_{12}\kappa(s_1, s_2; \mu_1, \mu_2) + \lambda_{13}\kappa(s_1, s_3; \mu_1, \mu_3) + \lambda_{23}\kappa(s_2, s_3; \mu_2, \mu_3) \\ &\quad +\lambda_1\kappa(s_1; \mu_1) + \lambda_2\kappa(s_2; \mu_2) + \lambda_3\kappa(s_3; \mu_3). \end{aligned}$$

Here the matrix $\boldsymbol{\Lambda}$ is defined by analogy with the bivariate case; see the general definition (17) below. This definition satisfies the requirement that each marginal distribution belongs to the univariate model $\mathrm{ED}^*(\mu, \lambda)$. For example, the CGF of the first margin is

$$\begin{aligned}
&K(s_1, 0, 0; \boldsymbol{\mu}, \boldsymbol{\Lambda}) \\
&\quad = \lambda_{12}\kappa(s_1, 0; \mu_1, \mu_2) + \lambda_{13}\kappa(s_1, 0; \mu_1, \mu_3) + \lambda_{23}\kappa(0, 0; \mu_2, \mu_3) \\
&\quad +\lambda_1\kappa(s_1; \mu_1) + \lambda_2\kappa(0; \mu_2) + \lambda_3\kappa(0; \mu_3) \\
&\quad = \lambda_{12}\kappa(s_1; \mu_1) + \lambda_{13}\kappa(s_1; \mu_1) + \lambda_1\kappa(s_1; \mu_1).
\end{aligned}$$

We also find that all three bivariate marginal distributions are of the form (10). For example, the CGF of the joint distribution of X_1 and X_2 is

$$\begin{aligned}
&K(s_1, s_2, 0; \boldsymbol{\mu}, \boldsymbol{\Lambda}) \\
&\quad = \lambda_{12}\kappa(s_1, s_2; \mu_1, \mu_2) + \lambda_{13}\kappa(s_1, 0; \mu_1, \mu_3) + \lambda_{23}\kappa(s_2, 0; \mu_2, \mu_3) \\
&\quad +\lambda_1\kappa(s_1; \mu_1) + \lambda_2\kappa(s_2; \mu_2) + \lambda_3\kappa(0; \mu_3) \\
&\quad = \lambda_{12}\kappa(s_1, s_2; \mu_1, \mu_2) + (\lambda_1 + \lambda_{13})\,\kappa(s_1; \mu_1) + (\lambda_2 + \lambda_{23})\,\kappa(s_2; \mu_2),
\end{aligned}$$

which has rate vector $(\mu_1, \mu_2)^\top$ and weight matrix given by the upper left 2×2 block of $\boldsymbol{\Lambda}$.

Based on these considerations, we now define the multivariate exponential dispersion model $\mathrm{ED}_k^*(\boldsymbol{\mu}, \boldsymbol{\Lambda})$ for general k by the following joint CGF:

$$K(\boldsymbol{s}; \boldsymbol{\mu}, \boldsymbol{\Lambda}) = \sum_{i<j} \lambda_{ij}\kappa(s_i, s_j; \mu_i, \mu_j) + \sum_{i=1}^{k} \lambda_i\kappa(s_i; \mu_i), \tag{16}$$

where all weights λ_{ij} and λ_i are positive. Defining

$$\lambda_{ii} = \sum_{\{j:j\neq i\}} \lambda_{ij} + \lambda_i, \tag{17}$$

the ith mean is easily seen to be $\mathrm{E}(X_i) = \lambda_{ii}\mu_i$, so that the mean vector takes the form

$$\mathrm{E}\,(\boldsymbol{X}) = \boldsymbol{\lambda\mu},$$

where $\boldsymbol{\Lambda} = \{\lambda_{ij}\}_{i,j=1}^{k}$, and $\boldsymbol{\lambda} = \mathrm{diag}\,\{\lambda_{11}, \ldots, \lambda_{kk}\}$ is a $k \times k$ diagonal matrix. By arguments similar to those given in the trivariate case above we find that each univariate margin belongs to the univariate model $\mathrm{ED}^*(\mu, \lambda)$, and that all marginal distributions for a subset of the k variables are again of the form (16).

Using (16) we find that $\mathrm{ED}_k^*(\boldsymbol{\mu}, \boldsymbol{\Lambda})$ satisfies the following (generalized) additive property:

$$\mathrm{ED}_k^*(\boldsymbol{\mu}, \boldsymbol{\Lambda}_1) + \mathrm{ED}_k^*(\boldsymbol{\mu}, \boldsymbol{\Lambda}_2) = \mathrm{ED}_k^*(\boldsymbol{\mu}, \boldsymbol{\Lambda}_1 + \boldsymbol{\Lambda}_2). \tag{18}$$

Table 1. Summary of ordinary and multivariate exponential dispersion models.

Form	Symbol	Mean vector	Covariance matrix
Additive	$\mathrm{ED}^*(\boldsymbol{\mu}, \lambda)$	$\lambda\boldsymbol{\mu}$	$\lambda \boldsymbol{V}(\boldsymbol{\mu})$
	$\mathrm{ED}_k^*(\boldsymbol{\mu}, \boldsymbol{\Lambda})$	$\boldsymbol{\lambda\mu}$	$\boldsymbol{\Lambda} \odot \boldsymbol{V}(\boldsymbol{\mu})$
Reproductive	$\mathrm{ED}(\boldsymbol{\mu}, \sigma^2)$	$\boldsymbol{\mu}$	$\sigma^2 \boldsymbol{V}(\boldsymbol{\mu})$
	$\mathrm{ED}_k(\boldsymbol{\mu}, \boldsymbol{\Sigma})$	$\boldsymbol{\mu}$	$\boldsymbol{\Sigma} \odot \boldsymbol{V}(\boldsymbol{\mu})$

Similar to the above discussion of the ordinary additive property (5), the property (18) gives rise to a stochastic process indexed by $\boldsymbol{\Lambda}$, having stationary and independent increments.

In order to calculate the covariance matrix, we let $V(\cdot)$ denote the unit variance function of the marginal distribution $\mathrm{ED}^*(\mu, \lambda)$, so that in particular the ith variance is

$$\mathrm{Var}(X_i) = \lambda_{ii} V(\mu_i).$$

From (16) we find that for $i \neq j$ the ijth covariance is

$$\mathrm{Cov}(X_i, X_j) = \lambda_{ij} V(\mu_i, \mu_j),$$

where $V(\mu_i, \mu_j) = \ddot{\kappa}(0, 0; \mu_i, \mu_j)$ denotes the second mixed derivative of $\kappa(\cdot, \cdot; \mu_1, \mu_2)$ at zero. Generalizing (13), we may hence express the covariance matrix for $\boldsymbol{X}$ as a Hadamard product,

$$\mathrm{Var}(\boldsymbol{X}) = \boldsymbol{\Lambda} \odot \boldsymbol{V}(\boldsymbol{\mu}), \tag{19}$$

where the (matrix) unit variance function $\mathbf{V}$ now has diagonal elements $V(\mu_i)$ and off-diagonal elements $V(\mu_i, \mu_j)$.

This construction gives us exactly the desired number of parameters, namely k rates and $k(k+1)/2$ covariance parameters. The use of the term *multivariate* exponential dispersion model hence signals that we have obtained a parallel to the *multivariate* normal distribution, in terms of the number of parameters and their interpretation. This will become clearer in Sec. 5 below, where we see that the multivariate normal distribution $\mathrm{N}_k(\boldsymbol{\mu}, \boldsymbol{\Sigma})$ is a multivariate reproductive exponential dispersion model. The multivariate normal distribution of the form $\mathrm{N}_k(\boldsymbol{\lambda\mu}, \boldsymbol{\Lambda})$ is an example of a multivariate additive exponential dispersion model. Table 1 summarizes the additive and reproductive forms of exponential dispersion models. The latter will be defined in Section 5 below.

Let us now consider the fully correlated case (14) (as opposed to the maximally correlated case where all $\lambda_i = 0$), which under the present conditions takes the form

$$V(\mu_i, \mu_j) = \pm\sqrt{V(\mu_i)V(\mu_j)}. \tag{20}$$

Choosing the positive sign, the covariance matrix (19) may now be expressed as follows:

$$\mathbf{\Lambda} \odot \boldsymbol{V}(\boldsymbol{\mu}) = \boldsymbol{V}^{1/2}(\boldsymbol{\mu})\mathbf{\Lambda}\boldsymbol{V}^{1/2}(\boldsymbol{\mu}),$$

where $\boldsymbol{V}^{1/2}(\boldsymbol{\mu}) = \operatorname{diag}\left[V^{1/2}(\mu_1), \ldots, V^{1/2}(\mu_k)\right]$. We shall return to this case in Sec. 5.

4. Multivariate discrete exponential dispersion models

4.1. *General*

We shall now consider the construction of multivariate versions of the Poisson, binomial and negative binomial distributions by the extended convolution method. Due to the additive property (18) the multivariate additive exponential dispersion model $\mathrm{ED}_k^*(\boldsymbol{\mu}, \mathbf{\Lambda})$ is well suited for discrete data, because the sum of two count variables is again a count variable. In the Poisson case we proceed by means of the ordinary convolution method, because, in spite of the problems mentioned above, this method turns out to give us the best possible solution in this case. In the binomial and negative binomial cases, however, we need to start from a suitable CGF (8) with correlated margins, in such a way that the correlation parameter is incorporated into the final model. These three examples hence give a good impression of the scope of the extended convolution method in the discrete case.

4.2. *A multivariate Poisson model*

In order to construct a multivariate Poisson model, we first consider the bivariate case, starting with the joint CGF defined by

$$\kappa(s_1, s_2) = e^{s_1+s_2} - 1 \text{ for } s_1, s_2 \in \mathbb{R}.$$

This CGF corresponds to the distribution of $(U, U)^\top$ with $U \sim \mathrm{Po}(1)$ (Poisson with mean 1), leading up to the ordinary convolution method. We hence obtain a bivariate Poisson distribution $\mathrm{Po}_2(\mathbf{\Lambda})$ via the joint CGF with positive weights λ_{12}, λ_1 and λ_2 given by

$$K(s_1, s_2; \mathbf{\Lambda}) = \lambda_{12}\left(e^{s_1+s_2} - 1\right) + \lambda_1\left(e^{s_1} - 1\right) + \lambda_2\left(e^{s_2} - 1\right),$$

see e.g. Holgate[7] and references therein. The two margins are univariate Poisson with means $\lambda_{11} = \lambda_{12} + \lambda_1$ and $\lambda_{22} = \lambda_{12} + \lambda_2$, respectively. The corresponding mean vector (rate vector) and covariance matrix (weight matrix) are

$$\begin{bmatrix}\lambda_{11}\\ \lambda_{22}\end{bmatrix} \text{ and } \mathbf{\Lambda} = \begin{bmatrix}\lambda_{11} & \lambda_{12}\\ \lambda_{12} & \lambda_{22}\end{bmatrix},$$

respectively, and hence the correlation between the two variables is

$$\rho = \frac{\lambda_{12}}{\sqrt{\lambda_{11}\lambda_{22}}}.$$

This correlation is positive and bounded by $\sqrt{\lambda_{\min}/\lambda_{\max}}$, involving the minimum and maximum of the two means λ_{11} and λ_{22}, see Holgate.[7]

For general $k \geq 2$ we define a multivariate Poisson distribution $\mathrm{Po}_k(\mathbf{\Lambda})$ by the joint CGF given by

$$K(\boldsymbol{s};\mathbf{\Lambda}) = \sum_{i<j} \lambda_{ij}\left(e^{s_i+s_j} - 1\right) + \sum_{i=1}^{k} \lambda_i\left(e^{s_i} - 1\right), \tag{21}$$

where all weights are positive. Again, the ith margin is Poisson with mean λ_{ii} defined by (17), and positive-definite covariance matrix $\mathbf{\Lambda} = \{\lambda_{ij}\}_{i,j=1}^{k}$. The additive property (18) takes the following form for the multivariate Poisson:

$$\mathrm{Po}_k(\mathbf{\Lambda}_1) + \mathrm{Po}_k(\mathbf{\Lambda}_2) = \mathrm{Po}_k(\mathbf{\Lambda}_1 + \mathbf{\Lambda}_2),$$

which generalizes the well-known convolution property of the univariate Poisson distribution.

Our multivariate Poisson distribution is a special case of the multivariate Poisson distribution of Teicher[8] and Dwass and Teicher.[9] These authors showed that the most general form of multivariate Poisson distribution obtainable by the convolution method has joint CGF of the following form (with all weights positive):

$$\sum_{i=1}^{k} \lambda_i\left(e^{s_i} - 1\right) + \sum_{i<j} \lambda_{ij}\left(e^{s_i+s_j} - 1\right) + \cdots + \lambda_{1\ldots k}\left(e^{s_1+\cdots+s_k} - 1\right), \tag{22}$$

which differs from (21) by including also terms of orders three up to k. This result implies that (21) is the most general form possible of the k-variate Poisson distribution involving only first- and second-order terms, a constraint that is crucial in order to parametrize the model by the covariance matrix $\mathbf{\Lambda}$.

4.3. *A multivariate binomial model*

There exist a number of different proposals for bivariate or multivariate binomial distributions, see e.g. Krishnamoorthy[10] and Marshall and Olkin.[11] Here we propose an extension of the bivariate binomial distributions introduced by Hamdan and Jensen[12] and Biswas and Hwang,[13] who both used a form of the extended convolution method in their constructions.

We begin with a pair of correlated Bernoulli variables, corresponding to the following 2×2 table of probabilities:

X_1		0	1
X_2	0	$\bar{\mu}_1\bar{\mu}_2 + \gamma$	$\mu_1\bar{\mu}_2 - \gamma$
	1	$\bar{\mu}_1\mu_2 - \gamma$	$\mu_1\mu_2 + \gamma$

Here $\mathrm{E}(X_i) = \mu_i = 1 - \bar{\mu}_i$ and $\mathrm{Var}(X_i) = \mu_i\bar{\mu}_i$ for $i = 1, 2$, and $\mathrm{Cov}(X_1, X_2) = \gamma$. To make all probabilities in the table positive, the covariance parameter γ must satisfy

$$-\min\{\mu_1\mu_2, \bar{\mu}_1\bar{\mu}_2\} \leq \gamma \leq \min\{\mu_1\bar{\mu}_2, \bar{\mu}_1\mu_2\}.$$

The bivariate joint CGF corresponding to the table has the form

$$\begin{aligned}\kappa(s_1, s_2; \mu_1, \mu_2, \gamma) = \log[1 &+ (\mu_1\mu_2 + \gamma)\left(e^{s_1+s_2} - 1\right) \\ &+ (\mu_1\bar{\mu}_2 - \gamma)\left(e^{s_1} - 1\right) + (\bar{\mu}_1\mu_2 - \gamma)\left(e^{s_2} - 1\right)],\end{aligned}$$

having both margins Bernoulli with CGFs $\kappa(s_i; \mu_i) = \log\left[1 + \mu_i\left(e^{s_i} - 1\right)\right]$ for $i = 1, 2$.

Starting once again with the bivariate case, we define the bivariate binomial $\mathrm{Bi}_2^\gamma(\boldsymbol{\mu}, \boldsymbol{N})$ by means of the joint CGF based on positive weights n_{12}, n_1 and n_2 given by

$$K(s_1, s_2; \boldsymbol{\mu}, \boldsymbol{N}, \gamma) = n_{12}\kappa(s_1, s_2; \mu_1, \mu_2, \gamma) + n_1\kappa(s_1; \mu_1) + n_2\kappa(s_2; \mu_2).$$

Here we denote the weights by n instead of λ in order to emphasize that the ns are all integers, due to the lack of infinite divisibility of the Bernoulli and binomial distributions. The symmetric matrix $\boldsymbol{N}$ is defined similar to $\boldsymbol{\Lambda}$ above. Following Hamdan and Jensen,[12] who studied the bivariate binomial distribution, we note that a vector $(X_1, X_2)^\top$ with distribution $\mathrm{Bi}_2^\gamma(\boldsymbol{\mu}, \boldsymbol{N})$ represents the total number of items of each of two types, where in n_{12} trials we classify the items according to both criteria, in n_1 trials we classify the items according to the first criterion only, and in n_2 trials we classify the items according to the second criterion only. The bivariate binomial distribution of Marshall and Olkin[11] is a special case of the bivariate binomial distribution $\mathrm{Bi}_2^\gamma(\boldsymbol{\mu}, \boldsymbol{N})$, obtained for $n_1 = n_2 = 0$.

Moving on to the general case, we define a multivariate binomial distribution $\mathrm{Bi}_k^{\gamma}(\boldsymbol{\mu}, \boldsymbol{N})$ by the following joint CGF:

$$K(\boldsymbol{s}; \boldsymbol{\mu}, \boldsymbol{N}, \boldsymbol{\gamma}) = \sum_{i<j} n_{ij} \kappa(s_i, s_j; \mu_i, \mu_j, \gamma_{ij}) + \sum_{i=1}^{k} n_i \log\left[1 + \mu_i\left(e^{s_i} - 1\right)\right], \tag{23}$$

where here and in the following $\boldsymbol{\gamma}$ denotes a vector of additional parameters (such as the covariances γ_{ij}) in addition to the rate and weight parameters. All margins are binomial, the ith having parameters (μ_i, n_{ii}), where

$$n_{ii} = \sum_{\{j:j\neq i\}} n_{ij} + n_i \text{ for } i = 1, \ldots, k,$$

similar to (17). Since we have started from a general bivariate Bernoulli distribution, this is the most general form of multivariate binomial distribution possible within our framework. The multivariate binomial distribution satisfies the following additive property:

$$\mathrm{Bi}_k^{\gamma}(\boldsymbol{\mu}, \boldsymbol{N}_1) + \mathrm{Bi}_k^{\gamma}(\boldsymbol{\mu}, \boldsymbol{N}_2) = \mathrm{Bi}_k^{\gamma}(\boldsymbol{\mu}, \boldsymbol{N}_1 + \boldsymbol{N}_2),$$

similar to the additive property of the univariate binomial distribution.

Let us consider the covariance structure of the multivariate binomial distribution. Since the margins are all binomial, the means and variances are

$$\mathrm{E}(X_i) = n_{ii}\mu_i \qquad \mathrm{Var}(X_i) = n_{ii}\mu_i\bar{\mu}_i \text{ for } i = 1, \ldots, k.$$

The ijth covariance is

$$\mathrm{Cov}(X_i, X_j) = n_{ij}\gamma_{ij} \text{ for } i \neq j, \tag{24}$$

corresponding to the correlation

$$\mathrm{Corr}(X_i, X_j) = \frac{n_{ij}}{\sqrt{n_{ii}n_{jj}}} \cdot \frac{\gamma_{ij}}{\sqrt{\mu_i\bar{\mu}_i\mu_j\bar{\mu}_j}},$$

whose sign is the same as that of γ_{ij}. We note that when the n_{ij} are known there are only $k(k-1)/2$ covariance parameters γ_{ij}, similar to the multivariate Poisson case above. These covariance parameters are of course inherited from the starting bivariate Bernoulli distribution, and do not derive from the weight parameters as in the general case.

4.4. *A multivariate negative binomial model*

To construct a multivariate negative binomial model we start from the bivariate negative multinomial model of Edwards and Gurland[14] and Subrahmaniam and Subrahmaniam.[15] This model has joint CGF proportional to the function

$$\kappa(s_1, s_2; \mu_1, \mu_2, \gamma) = \\ -\log\left[1 - \gamma\left(e^{s_1+s_2} - 1\right) - (\mu_1 - \gamma)\left(e^{s_1} - 1\right) - (\mu_2 - \gamma)\left(e^{s_2} - 1\right)\right]. \quad (25)$$

The joint CGF (25) is parametrized by the marginal means μ_1 and μ_2 (both positive) and the covariance parameter $\gamma \in [0, \min\{\mu_1, \mu_2\})$. Both margins are geometric, corresponding to the CGF $\kappa(s_i; \mu_i) = -\log\left[1 - \mu_i\left(e^{s_i} - 1\right)\right]$ for $i = 1, 2$, and the two variances are $\mathrm{Var}(X_i) = \mu_i(1 + \mu_i)$ for $i = 1, 2$. The covariance is $\mathrm{Cov}(X_1, X_2) = \gamma + \mu_1\mu_2$.

We now define the multivariate negative binomial distribution generated from (25) by the following joint CGF with positive weights λ_{ij} and λ_i:

$$K(\boldsymbol{s}; \boldsymbol{\mu}, \boldsymbol{\Lambda}, \gamma) = \sum_{i<j} \lambda_{ij}\kappa(s_i, s_j; \mu_i, \mu_j, \gamma_{ij}) - \sum_{i=1}^{k} \lambda_i \log\left[1 - \mu_i\left(e^{s_i} - 1\right)\right]. \quad (26)$$

The margins are all negative binomial with means $\lambda_{ii}\mu_i$ and variances $\lambda_{ii}\mu_i(1 + \mu_i)$, where λ_{ii} is defined by (17). In the special case where all $\lambda_{ii} = 1$ we obtain a multivariate geometric distribution. The ijth covariance is

$$\mathrm{Cov}(X_i, X_j) = \lambda_{ij}\left(\gamma_{ij} + \mu_i\mu_j\right) \text{ for } i \neq j, \quad (27)$$

and the corresponding correlation is

$$\mathrm{Corr}(X_i, X_j) = \frac{\lambda_{ij}}{\sqrt{\lambda_{ii}\lambda_{jj}}} \cdot \frac{\gamma_{ij} + \mu_i\mu_j}{\sqrt{\mu_i\left(1 + \mu_i\right)\mu_j\left(1 + \mu_j\right)}}.$$

This correlation is positive and bounded by

$$\sqrt{\mu_{\min}\left(1 + \mu_{\min}\right)/\mu_{\max}\left(1 + \mu_{\max}\right)},$$

involving the minimum and maximum of μ_i and μ_j, similar to the Poisson case above.

By comparison, Doss[16] proposed a multivariate generalization of (25) based on the joint CGF

$$K(\boldsymbol{s}) = -\lambda \log\left[\sum a_{i_1 \ldots i_r} e^{s_{i_1} + \cdots + s_{i_r}}\right],$$

where the summation is over all subsets $\{i_1, \ldots, i_r\}$ of the set $\{1, \ldots, k\}$. This model, which is different from (26), may be obtained as a gamma mixture of the multivariate Poisson model (22).

The above construction illustrates the case where we base our construction on a specific bivariate negative multinomial model which, similar to the binomial case, has a covariance parameter that is inherited by the multivariate generalization. We observe from (27) that the covariance parameter γ_{ij} is confounded with the weight parameter λ_{ij}.

5. Convolution method, reproductive form

5.1. *General*

We shall now introduce multivariate exponential dispersion models of reproductive form, which we do by means of a generalization of the ordinary duality transformation (6). To this end we use the $k \times k$ diagonal matrix $\boldsymbol{\lambda} = \operatorname{diag}\{\boldsymbol{\Lambda}\}$ from above (containing the diagonal of $\boldsymbol{\Lambda}$), and define the reproductive multivariate exponential dispersion model $\mathrm{ED}_k(\boldsymbol{\mu}, \boldsymbol{\Sigma})$ by the (extended) duality transformation

$$\mathrm{ED}_k(\boldsymbol{\mu}, \boldsymbol{\Sigma}) = \boldsymbol{\lambda}^{-1}\mathrm{ED}_k^*(\boldsymbol{\mu}, \boldsymbol{\Lambda}). \tag{28}$$

The reproductive model $\mathrm{ED}_k(\boldsymbol{\mu}, \boldsymbol{\Sigma})$ has mean vector $\boldsymbol{\mu}$ and *dispersion matrix* $\boldsymbol{\Sigma} = \boldsymbol{\lambda}^{-1}\boldsymbol{\Lambda}\boldsymbol{\lambda}^{-1}$. It satisfies the following reproductive property, generalizing (7), namely that for $\boldsymbol{Y}_1, \ldots, \boldsymbol{Y}_n$ i.i.d. $\mathrm{ED}_k(\boldsymbol{\mu}, \boldsymbol{\Sigma})$,

$$\frac{1}{n}\sum_{i=1}^{n} \boldsymbol{Y}_i \sim \mathrm{ED}_k(\boldsymbol{\mu}, \boldsymbol{\Sigma}/n).$$

Just as for the additive form there are k means and $k(k+1)/2$ dispersion parameters. The additive and reproductive forms of multivariate exponential dispersion models are summarized in Table 1 at the end of Sec. 3.

The covariance matrix for $\mathrm{ED}_k(\boldsymbol{\mu}, \boldsymbol{\Sigma})$ has the form $\boldsymbol{\Sigma} \odot \boldsymbol{V}(\boldsymbol{\mu})$, e.g. for $k = 2$:

$$\boldsymbol{\Sigma} \odot \boldsymbol{V}(\boldsymbol{\mu}) = \begin{bmatrix} \sigma_{11}V_{11}(\boldsymbol{\mu}) & \sigma_{12}V_{12}(\boldsymbol{\mu}) \\ \sigma_{21}V_{21}(\boldsymbol{\mu}) & \sigma_{22}V_{22}(\boldsymbol{\mu}) \end{bmatrix} = \begin{bmatrix} \frac{1}{\lambda_{11}}V_{11}(\boldsymbol{\mu}) & \frac{\lambda_{12}}{\lambda_{11}\lambda_{22}}V_{12}(\boldsymbol{\mu}) \\ \frac{\lambda_{12}}{\lambda_{11}\lambda_{22}}V_{21}(\boldsymbol{\mu}) & \frac{1}{\lambda_{22}}V_{22}(\boldsymbol{\mu}) \end{bmatrix}.$$

In particular the correlation is the same as in the additive form, namely

$$\frac{\sigma_{12}}{\sqrt{\sigma_{11}\sigma_{22}}}\frac{V_{12}(\boldsymbol{\mu})}{\sqrt{V_{11}(\boldsymbol{\mu})V_{22}(\boldsymbol{\mu})}} = \frac{\lambda_{12}}{\sqrt{\lambda_{11}\lambda_{22}}}\frac{V_{12}(\boldsymbol{\mu})}{\sqrt{V_{11}(\boldsymbol{\mu})V_{22}(\boldsymbol{\mu})}},$$

and the same bounds apply as those discussed in Sec. 3.1.

According to the duality transformation (28), each additive exponential dispersion model $\mathrm{ED}_k^*(\boldsymbol{\mu}, \boldsymbol{\Lambda})$ has a corresponding reproductive counterpart. The inverse duality transformation is given by

$$\mathrm{ED}_k^*(\boldsymbol{\mu}, \boldsymbol{\Lambda}) = \boldsymbol{\lambda}\mathrm{ED}_k(\boldsymbol{\mu}, \boldsymbol{\lambda}^{-1}\boldsymbol{\Lambda}\boldsymbol{\lambda}^{-1}), \tag{29}$$

by which the additive form $\mathrm{ED}_k^*(\boldsymbol{\mu}, \boldsymbol{\Lambda})$ may be recovered from the reproductive form.

The multivariate normal distribution $\mathrm{N}_k(\boldsymbol{\mu}, \boldsymbol{\Sigma})$ is the main example of a multivariate reproductive exponential dispersion model. In general, the properties of the reproductive exponential dispersion model $\mathrm{ED}_k(\boldsymbol{\mu}, \boldsymbol{\Sigma})$ make it particularly suitable for continuous data that represent yield per unit of volume or similar.

5.2. *A multivariate gamma model*

We now expand the bivariate gamma model of Jørgensen[1] into a multivariate gamma exponential dispersion model, generalizing the multivariate gamma of Mathai and Moschopoulos.[17] Starting from a unit exponential variable U, we consider the bivariate random vector

$$\begin{bmatrix} \mu_1 U \\ \mu_2 U \end{bmatrix}, \tag{30}$$

with mean vector $\boldsymbol{\mu} = (\mu_1, \mu_2)^\top \in \mathbb{R}^2$ and joint CGF

$$\kappa(s_1, s_2; \mu_1, \mu_2) = -\log(1 - \mu_1 s_1 - \mu_2 s_2). \tag{31}$$

This is a good starting point for our construction, because the margins of (30) are fully correlated and yet have two independent mean parameters. We hence define a multivariate gamma distribution of additive form, denoted $\mathrm{Ga}_k^*(\boldsymbol{\mu}, \boldsymbol{\Lambda})$, by means of the joint CGF

$$K(\boldsymbol{s}; \boldsymbol{\mu}, \boldsymbol{\Lambda}) = -\sum_{i<j} \lambda_{ij} \log(1 - \mu_i s_i - \mu_j s_j) - \sum_{i=1}^{k} \lambda_i \log(1 - \mu_i s_i). \tag{32}$$

This is a special case of the multivariate gamma distribution of Prékopa and Szántai,[18] the difference being that these authors included terms of order three and higher, much like the multivariate Poisson distribution (22), in addition to the first- and second-order terms used in (32). In the special case where $\lambda_{ii} = 1$ for all $i = 1, \ldots, k$ we obtain a multivariate exponential distribution.

By means of the duality transformation (28) we obtain the reproductive form $\mathrm{Ga}_k(\boldsymbol{\mu}, \boldsymbol{\Sigma})$ with mean vector $\boldsymbol{\mu}$ and dispersion matrix $\boldsymbol{\Sigma}$ with entries σ_{ij}. Here the unit variance function $\boldsymbol{V}(\boldsymbol{\mu})$ has entries defined by

$$V_{ij}(\boldsymbol{\mu}) = \mu_i \mu_j = \sqrt{V_{ii}(\boldsymbol{\mu}) V_{jj}(\boldsymbol{\mu})}$$

for $i, j = 1, \ldots, k$, in line with (20). The corresponding covariance matrix is

$$\boldsymbol{\Sigma} \odot \boldsymbol{V}(\boldsymbol{\mu}) = \mathrm{diag}(\boldsymbol{\mu}) \boldsymbol{\Sigma} \mathrm{diag}(\boldsymbol{\mu}), \tag{33}$$

and the ijth correlation is

$$\mathrm{Corr}(X_i, X_j) = \frac{\sigma_{ij}}{\sqrt{\sigma_{ii}\sigma_{jj}}},$$

which varies between 0 and 1, in line with the above discussion of the condition (14). In the present case the extended convolution method hence interpolates between the set of fully correlated bivariate gamma models corresponding to the first term of (32), and the independent margins corresponding to the second term of (32).

The reproductive multivariate gamma distribution $\mathrm{Ga}_k(\boldsymbol{\mu}, \boldsymbol{\Sigma})$ satisfies the following scaling property:

$$\boldsymbol{C}\mathrm{Ga}_k(\boldsymbol{\mu}, \boldsymbol{\Sigma}) = \mathrm{Ga}_k(\boldsymbol{C}\boldsymbol{\mu}, \boldsymbol{\Sigma}), \tag{34}$$

where $\boldsymbol{C}$ is a diagonal matrix with positive entries. By combining this with the inverse duality transformation (29), we find that the additive form of the multivariate gamma may be expressed as follows:

$$\mathrm{Ga}_k^*(\boldsymbol{\mu}, \boldsymbol{\Lambda}) = \mathrm{Ga}_k(\boldsymbol{\lambda}\boldsymbol{\mu}, \boldsymbol{\lambda}^{-1}\boldsymbol{\Lambda}\boldsymbol{\lambda}^{-1}), \tag{35}$$

which implies that the additive and reproductive forms of the multivariate gamma are equivalent up to a reparametrization.

There is a large number of different bivariate and multivariate gamma distributions available in the literature. For example, Balakrishnan and Lai,[5] Ch. 8, list more than twenty-five different types of bivariate gamma distributions. Our particular construction is designed to give the maximum possible domain for the correlation, and in addition it yields an intuitively appealing form of covariance matrix. The construction is canonical in the sense that its starting point is the singular distribution (30) based on a single unit exponential variable U.

6. Multivariate Tweedie models

We shall now introduce a new class of multivariate Tweedie distributions using a construction similar to the method used for the multivariate gamma distribution of Sec. 5.2. First we review the univariate Tweedie models, following Jørgensen,[2] Ch. 4.

6.1. *The univariate case*

The reproductive form of the univariate Tweedie distributions is characterized by having power unit variance functions

$$V(\mu) = \mu^p, \tag{36}$$

where the power parameter p has domain $\Delta = (-\infty, 0] \cup [1, \infty)$. The domain for μ is $\mathbb{R}$ for $p = 0$ and $\mathbb{R}_+$ for all other values of p. The reproductive Tweedie exponential dispersion model corresponding to (36) is denoted $\mathrm{Tw}^p(\mu, \sigma^2)$, and has mean μ and variance $\sigma^2\mu^p$, where $\sigma^2 > 0$. This model satisfies the scaling property

$$c\mathrm{Tw}^p(\mu, \sigma^2) = \mathrm{Tw}^p(c\mu, c^{2-p}\sigma^2), \tag{37}$$

which is closely related to the Tweedie convergence theorem of Jørgensen, Martnez and Tsao,[19] see also Jørgensen,[2] Ch. 4. The Poisson and gamma distributions are both Tweedie models, so that $\mathrm{Tw}^1(\mu, 1) = \mathrm{Po}(\mu)$ and $\mathrm{Tw}^2(\mu, \sigma^2) = \mathrm{Ga}(\mu, \sigma^2)$. The inverse duality transformation in combination with (37) lead to the following additive form of the Tweedie model:

$$\mathrm{Tw}^{*p}(\mu, \lambda) = \lambda\mathrm{Tw}^p(\mu, \lambda^{-1}) = \mathrm{Tw}^p(\lambda\mu, \lambda^{1-p}).$$

Just like in the gamma case, we find that the additive and reproductive forms of the Tweedie model are equivalent up to a reparametrization.

To prepare for the multivariate construction, we now consider the Tweedie CGF. To this end, we first introduce the parameter $\alpha = 1 + (1 - p)^{-1}$ with domain $[-\infty, 1) \cup (1, 2]$ (Jørgensen,[2] p. 131). Setting aside the gamma and Poisson cases, let us consider a $p \in \Delta \setminus \{1, 2\}$, corresponding to $\alpha \neq -\infty, 0$. In this case the Tweedie cumulant function κ_α is defined by

$$\kappa_\alpha(\theta) = \frac{\alpha - 1}{\alpha}\left(\frac{\theta}{\alpha - 1}\right)^\alpha$$

for values of θ such that $\theta/(\alpha - 1) > 0$. The additive Tweedie model $\mathrm{Tw}^{*p}(\mu, \lambda)$ has CGF

$$s \longmapsto \lambda\left[\kappa_\alpha(s + \theta) - \kappa_\alpha(\theta)\right] = \lambda\kappa_\alpha(\theta)\left[\left(1 + \frac{s}{\theta}\right)^\alpha - 1\right], \tag{38}$$

for values of s such that $s/\theta > -1$. The parameters μ and θ are related by

$$\mu = \dot{\kappa}_\alpha(\theta) = \left(\frac{\theta}{\alpha - 1}\right)^{\alpha-1}. \tag{39}$$

In the case of the normal distribution ($p = 0$, $\alpha = 2$), both μ and θ have domain $\mathbb{R}$.

6.2. *The multivariate construction*

Let α correspond to a value of p in $\Delta \setminus \{1, 2\}$ as above. In order to define the multivariate Tweedie model, we introduce the intermediate weight parameter $\gamma = \lambda\kappa_\alpha(\theta)$, and write the CGF (38) as follows:

$$s \longmapsto \gamma\left[\left(1 + \frac{s}{\theta}\right)^\alpha - 1\right]. \tag{40}$$

Since γ and $\kappa_\alpha(\theta)$ have the same sign, it follows that the domain for γ is either $\mathbb{R}_+$ or $\mathbb{R}_-$, depending on the sign of $(\alpha - 1)/\alpha$. Our starting point is the bivariate singular distribution with joint CGF

$$(s_1, s_2)^\top \longmapsto \gamma\left[\left(1 + \frac{s_1}{\theta_1} + \frac{s_2}{\theta_2}\right)^\alpha - 1\right],$$

whose margins are Tweedie distributions of the form (40), similar to the bivariate singular gamma distribution (31). We define the multivariate additive Tweedie model by the joint CGF

$$K(\boldsymbol{s}; \boldsymbol{\theta}, \boldsymbol{\gamma}) = \sum_{i<j} \gamma_{ij}\left[\left(1 + \frac{s_i}{\theta_i} + \frac{s_j}{\theta_j}\right)^\alpha - 1\right] + \sum_{i=1}^{k} \gamma_i\left[\left(1 + \frac{s_i}{\theta_i}\right)^\alpha - 1\right], \tag{41}$$

where the weight parameters γ_{ij} and γ_i all have the same sign as $(\alpha - 1)/\alpha$. By taking $s_j = 0$ for $j \neq i$ in the expression (41) we find that the ith margin follows a univariate Tweedie distribution with CGF (40) with $\theta = \theta_i$ and $\gamma = \gamma_{ii}$ defined by

$$\gamma_{ii} = \sum_{\{j:j\neq i\}} \gamma_{ij} + \gamma_i.$$

We have hence obtained a multivariate Tweedie distribution with the required number of parameters, except that we need to define a parametrization similar to the univariate case given by (38) and (39).

We hence define the weight parameters $\lambda_{ii}, \lambda_{ij} > 0$ by

$$\lambda_{ii} = \frac{\gamma_{ii}}{\kappa_\alpha(\theta_i)} \text{ for } i = 1, \ldots, k \tag{42}$$

and

$$\lambda_{ij} = \frac{\gamma_{ij}}{\kappa_\alpha^{1/2}(\theta_i, \theta_j)} \text{ for } i < j, \tag{43}$$

where $\kappa_\alpha^{1/2}$ is the function defined by

$$\kappa_\alpha^{1/2}(\theta_i, \theta_j) = \frac{\alpha - 1}{\alpha} \left(\frac{\theta_i}{\alpha - 1} \right)^{\alpha/2} \left(\frac{\theta_j}{\alpha - 1} \right)^{\alpha/2}.$$

Using the parameters λ_{ii} the marginal means are of the form

$$\lambda_{ii}\mu_i = \lambda_{ii}\kappa_\alpha(\theta_i)\frac{\alpha}{\theta_i} = \lambda_{ii} \left(\frac{\theta_i}{\alpha - 1} \right)^{\alpha - 1} \text{ for } i = 1, \ldots, k, \tag{44}$$

consistent with (39). The variances are

$$\lambda_{ii}\kappa_\alpha(\theta_i)\frac{\alpha(\alpha - 1)}{\theta_i^2} = \lambda_{ii}\mu_i^p \text{ for } i = 1, \ldots, k,$$

which is consistent with (36). This defines the multivariate additive Tweedie model, denoted $\boldsymbol{X} \sim \mathrm{Tw}_k^{*p}(\boldsymbol{\mu}, \boldsymbol{\Lambda})$, where the elements of $\boldsymbol{\mu}$ are defined by (44), and $\boldsymbol{\Lambda}$ has elements λ_{ij} as above. The ith margin of $\boldsymbol{X}$ is univariate Tweedie, $X_i \sim \mathrm{Tw}^{*p}(\mu_i, \lambda_{ii})$.

The covariance between two components of $\boldsymbol{X}$ is

$$\mathrm{Cov}(X_i, X_j) = \lambda_{ij}\kappa_\alpha^{1/2}(\theta_i, \theta_j)\frac{\alpha(\alpha - 1)}{\theta_i\theta_j} = \lambda_{ij}(\mu_i\mu_j)^{p/2}.$$

The corresponding correlation is

$$\mathrm{Corr}(X_i, X_j) = \frac{\lambda_{ij}(\mu_i\mu_j)^{p/2}}{\sqrt{\lambda_{ii}\mu_i^p\lambda_{jj}\mu_j^p}} = \frac{\lambda_{ij}}{\sqrt{\lambda_{ii}\lambda_{jj}}} = \frac{|\gamma_{ij}|}{\sqrt{\gamma_{ii}\gamma_{jj}}},$$

which is free of μ and varies between 0 and 1. The covariance matrix for $\boldsymbol{X}$ has the form $\boldsymbol{\Lambda} \odot \boldsymbol{V}(\boldsymbol{\mu})$, where $\boldsymbol{V}(\boldsymbol{\mu})$ has entries

$$V_{ij}(\boldsymbol{\mu}) = (\mu_i\mu_j)^{p/2} = \sqrt{V_{ii}(\boldsymbol{\mu})V_{jj}(\boldsymbol{\mu})}. \tag{45}$$

The multivariate additive Tweedie model $\mathrm{Tw}_k^{*p}(\boldsymbol{\mu}, \boldsymbol{\Lambda})$ satisfies the following additive property:

$$\mathrm{Tw}_k^{*p}(\boldsymbol{\mu}, \boldsymbol{\Lambda}_1) + \mathrm{Tw}_k^{*p}(\boldsymbol{\mu}, \boldsymbol{\Lambda}_2) = \mathrm{Tw}_k^{*p}(\boldsymbol{\mu}, \boldsymbol{\Lambda}_1 + \boldsymbol{\Lambda}_2).$$

This result follows by noting that, for given value of $\boldsymbol{\theta}$, convolution of different members of (41) corresponds to adding the two values of γ_{ij}, which in turn, by (42) and (43), corresponds to adding the two values of λ_{ij}.

Table 2. Summary of multivariate Tweedie dispersion models.

Case	Symbol	Power
Extreme stable	$\mathrm{Tw}_k^p(\boldsymbol{\mu}, \boldsymbol{\Sigma})$	$p < 0$
Normal	$\mathrm{N}_k(\boldsymbol{\mu}, \boldsymbol{\Sigma})$	$p = 0$
Poisson	$\mathrm{Po}_k(\boldsymbol{\Lambda})$	$p = 1$
Compound Poisson	$\mathrm{Tw}_k^p(\boldsymbol{\mu}, \boldsymbol{\Sigma})$	$1 < p < 2$
Gamma	$\mathrm{Ga}_k(\boldsymbol{\mu}, \boldsymbol{\Sigma})$	$p = 2$
Positive stable	$\mathrm{Tw}_k^p(\boldsymbol{\mu}, \boldsymbol{\Sigma})$	$p > 2$
Inverse Gaussian	$\mathrm{IG}_k(\boldsymbol{\mu}, \boldsymbol{\Sigma})$	$p = 3$

We may now define the reproductive form $\mathrm{Tw}_k^p(\boldsymbol{\mu}, \boldsymbol{\Sigma})$ of the multivariate Tweedie model by means of the duality transformation

$$\mathrm{Tw}_k^p(\boldsymbol{\mu}, \boldsymbol{\Sigma}) = \boldsymbol{\lambda}^{-1}\mathrm{Tw}_k^{*p}(\boldsymbol{\mu}, \boldsymbol{\Lambda}).$$

The distribution $\mathrm{Tw}_k^p(\boldsymbol{\mu}, \boldsymbol{\Sigma})$ has mean vector $\boldsymbol{\mu}$ and dispersion matrix $\boldsymbol{\Sigma} = \boldsymbol{\lambda}^{-1}\boldsymbol{\Lambda}\boldsymbol{\lambda}^{-1}$ as above. The covariance matrix is

$$\boldsymbol{\Sigma} \odot \boldsymbol{V}(\boldsymbol{\mu}) = \mathrm{diag}(\boldsymbol{\mu})^{p/2}\boldsymbol{\Sigma}\mathrm{diag}(\boldsymbol{\mu})^{p/2}, \tag{46}$$

which is of sandwich form, similar to the gamma case (33). The different cases of multivariate Tweedie distributions are summarized in Table 2, including the multivariate normal, Poisson and gamma models. The multivariate inverse Gaussian distribution in the table refers to the multivariate reproductive Tweedie model $\mathrm{Tw}_k^3(\boldsymbol{\mu}, \boldsymbol{\Sigma})$, whose margins are univariate inverse Gaussian.

The reproductive Tweedie model $\mathrm{Tw}_k^p(\boldsymbol{\mu}, \boldsymbol{\Sigma})$ satisfies the following scaling property

$$\boldsymbol{C}\mathrm{Tw}_k^p(\boldsymbol{\mu}, \boldsymbol{\Sigma}) = \mathrm{Tw}_k^p(\boldsymbol{C}\boldsymbol{\mu}, \boldsymbol{C}^{1-p/2}\boldsymbol{\Sigma}\boldsymbol{C}^{1-p/2}), \tag{47}$$

where $\boldsymbol{C}$ is a diagonal matrix with positive entries, generalizing (34) and (37). By means of the scaling property, we may write the inverse duality transformation as follows:

$$\mathrm{Tw}_k^{*p}(\boldsymbol{\mu}, \boldsymbol{\Lambda}) = \boldsymbol{\lambda}\mathrm{Tw}_k^p(\boldsymbol{\mu}, \boldsymbol{\lambda}^{-1}\boldsymbol{\Lambda}\boldsymbol{\lambda}^{-1}) = \mathrm{Tw}_k^p(\boldsymbol{\lambda}\boldsymbol{\mu}, \boldsymbol{\lambda}^{-p/2}\boldsymbol{\Lambda}\boldsymbol{\lambda}^{-p/2}),$$

generalizing the form (35) for the gamma distribution.

We note that the scaling property (47) may be rewritten as a fixed point:

$$\boldsymbol{C}^{-1}\mathrm{Tw}_k^p\left(\boldsymbol{C}\boldsymbol{\mu}, \boldsymbol{C}^{1-p/2}\boldsymbol{\Sigma}\boldsymbol{C}^{1-p/2}\right) = \mathrm{Tw}_k^p(\boldsymbol{\mu}, \boldsymbol{\Sigma}),$$

compare with Jørgensen, Martnez and Demtrio.[20] In the univariate case, convergence to such a fixed point was explored by Jørgensen, Martnez and

Tsao[19] and Jørgensen, Martnez and Vinogradov,[21] which leads us to conjecture that a similar convergence result holds in the multivariate case. If $\mathrm{ED}_k(\boldsymbol{\mu}, \boldsymbol{\Sigma})$ denotes a multivariate exponential dispersion model based on a univariate unit variance function with power asymptotic behaviour $V(\mu) \sim \mu^p$, we hence conjecture that a convergence result of the following form applies:

$$\boldsymbol{C}^{-1}\mathrm{ED}_k(\boldsymbol{C}\boldsymbol{\mu}, \boldsymbol{C}^{1-p/2}\boldsymbol{\Sigma}\boldsymbol{C}^{1-p/2}) \xrightarrow{d} \mathrm{Tw}_k^p(\boldsymbol{\mu}, \boldsymbol{\Sigma}),$$

as $\boldsymbol{C} \to \infty$ in a suitable sense, with $\xrightarrow{d}$ denoting convergence in distribution. If correct, this conjecture implies that the multivariate Tweedie model, like in the univariate case, can be expected to occur frequently in practice. Further details about this conjecture will be published elsewhere.

Other forms of multivariate Tweedie models have been proposed recently by Furman[22] and Grigelionis.[23] The multivariate Tweedie model of Furman[22] is somewhat similar to the multivariate gamma model of Mathai,[17] and shares with the latter the property that is has only $2(k+1)$ shape and scale parameters, which in general falls short of the $k+k(k+1)/2$ parameters that we require. The multivariate Tweedie models of Grigelionis[23] are defined via a multivariate Lévy-Khinchine representation, and do not have univariate margins of Tweedie form.

7. Multivariate generalized linear models

We have now developed the main techniques for constructing multivariate exponential dispersion model, along with several important examples. Based on these results we now propose a new class of multivariate generalized linear models. We shall not develop this topic in full generality here, but it is useful to see how the basic principles of univariate generalized linear models can be extended to the multivariate case.

A univariate generalized linear model combines two basic ingredients, namely an error distribution and a link function, allowing the mean of the response variable to be modelled as a function of covariates. In the case of a vector of response variables, we generalize the classical multivariate multiple linear regression models to allow for non-normal errors and a common non-linear link function, say.

Let us consider independent k-variate response vectors $\boldsymbol{Y}_1, \ldots, \boldsymbol{Y}_n$, and let us first assume a multivariate reproductive Tweedie error distribution

$$\boldsymbol{Y}_i \sim \mathrm{Tw}_k^p(\boldsymbol{\mu}_i, \boldsymbol{\Sigma}), \tag{48}$$

where the power parameter p is considered known. For a given link function g, we consider the regression model defined by $g(\boldsymbol{\mu}_i) = \boldsymbol{B}\boldsymbol{x}_i$, where g maps the mean vector $\boldsymbol{\mu}_i$ elementwise to the k-variate linear predictor $\boldsymbol{B}\boldsymbol{x}_i$, and where $\boldsymbol{B}$ is a $k \times \ell$ matrix of regression coefficients and $\boldsymbol{x}_i$ is an ℓ-vector of covariates. In the special case where g is the identity function and $\mathrm{Tw}_k^p(\boldsymbol{\mu}, \boldsymbol{\Sigma})$ is the multivariate normal distribution $\mathrm{N}_k(\boldsymbol{\mu}_i, \boldsymbol{\Sigma})$, we obtain the classical multivariate multiple linear regression model. In the special case $p = 2$ we obtain a multivariate gamma regression model.

Taking advantage of the sandwich form (46) of the covariance matrix in the Tweedie case we propose the following $k \times \ell$ quasi-score function for estimating the regression matrix $\boldsymbol{B}$:

$$\boldsymbol{\psi}(\boldsymbol{B}) = \sum_{i=1}^{n} \left[\dot{g}(\boldsymbol{\mu}_i)\boldsymbol{V}(\boldsymbol{\mu}_i)\right]^{-1} (\boldsymbol{Y}_i - \boldsymbol{\mu}_i)\, \boldsymbol{x}_i^\top,$$

where $\boldsymbol{V}$ is defined by (45) and $\dot{g}(\boldsymbol{\mu}_i)$ is the $k \times k$ diagonal matrix of derivatives of the link function. Solving the quasi-score equation $\boldsymbol{\psi}(\boldsymbol{B}) = \boldsymbol{0}$ corresponds to fitting k separate generalized linear models to the entrics of $\boldsymbol{Y}_i$, in much the same way that a multivariate multiple linear regression model may be fitted by means of k separate multiple linear regressions. We let $\widehat{\boldsymbol{B}}$ denote the quasi-score estimator, and we let $\widehat{\boldsymbol{\mu}}_i = g^{-1}\left(\widehat{\boldsymbol{B}}\boldsymbol{x}_i\right)$ denote the corresponding vectors of fitted values.

Under the multivariate Tweedie model (48) we obtain

$$\boldsymbol{\Sigma} = \mathrm{Cov}\left[\boldsymbol{V}^{-1/2}(\boldsymbol{\mu}_i)\boldsymbol{Y}_i\right],$$

which allows us to estimate $\boldsymbol{\Sigma}$ by means of the following Pearson estimator:

$$\widehat{\boldsymbol{\Sigma}} = \frac{1}{n-\ell}\sum_{i=1}^{n} \boldsymbol{V}^{-1/2}(\widehat{\boldsymbol{\mu}}_i)\,(\boldsymbol{Y}_i - \widehat{\boldsymbol{\mu}}_i)\,(\boldsymbol{Y}_i - \widehat{\boldsymbol{\mu}}_i)^\top\, \boldsymbol{V}^{-1/2}(\widehat{\boldsymbol{\mu}}_i),$$

which generalizes the estimator based on the sum of squares and cross-products of residuals known from multivariate multiple linear regression.

Finally, let us give a few pointers to regression modelling for multivariate discrete exponential dispersion models. In the Poisson case, we regress the vector of means on $\boldsymbol{x}_i$ as above, with unit variance function $\boldsymbol{V}$ defined by (45) with $p = 1$. The off-diagonal elements of the variance matrix $\boldsymbol{\Lambda}$ may then be estimated by the empirical covariance matrix for the residuals. In the binomial and negative binomial cases we estimate the regression parameters in a similar way, and then estimate the covariance parameters γ_{ij} by equating the covariances (24) or (27) to the empirical covariance for the residuals.

8. Discussion

We have developed a unified methodology for the construction of multivariate exponential dispersion models, leading to multivariate versions of a range of common distributions. By extending the well-known convolution method, we obtain multivariate models with a fully flexible correlation structure, in terms of the number of covariance parameters of the models. Our ultimate goal is to produce a streamlined methodology for regression modelling of multivariate non-normal data, via a multivariate version of Nelder and Wedderburn's[24] generalized linear models. The multivariate exponential dispersion models proposed here have all margins of the same form, but it seems plausible that methods for combining different types of response variables (e.g. discrete and continuous) can be developed.

Many classical normal-theory models, such as time series, variance components and graphical models, are based on imposing certain constraints on the covariance matrix of the multivariate normal distribution. Analogues of such models may easily be constructed by imposing similar constraints on the dispersion matrix for a given multivariate dispersion model. In a similar vein, it would seem possible to develop analogues of many conventional multivariate techniques such as principal components or canonical correlation analysis.

One limitation of the extended convolution method is that the pairwise correlation is restricted to positive values, often with an upper bound. A further limitation is that the probability density function generally involves an intractable high-dimensional integral or sum, making the calculation of probabilities for multivariate exponential dispersion models computationally heavy. This also complicates the use of likelihood methods for statistical inference, although this obstacle can be overcome by using quasi-likelihood methods.

As shown by Jørgensen,[1] principles similar to those discussed here apply to the construction of other types of multivariate dispersion models, such as proper and extreme dispersion models. Table 3 gives an overview of the different kinds of ordinary and multivariate dispersion models that have been proposed so far. Many details of these models remain to be explored, including a possible multivariate generalization of the class of geometric dispersion models of Jørgensen and Kokonendji.[25] A website[26] contains further information about ongoing research on multivariate dispersion models.

Table 3. Summary of different types of ordinary and multivariate dispersion models.

Type	Ordinary	Multivariate	References
Proper	$\mathrm{PD}(\mu, \sigma^2)$	$\mathrm{PD}_k(\boldsymbol{\mu}, \boldsymbol{\Sigma})$	Jørgensen,[2] Ch. 5 Jørgensen and Lauritzen[3]
Exponential	$\mathrm{ED}^*(\mu, \lambda)$ $\mathrm{ED}(\mu, \sigma^2)$	$\mathrm{ED}_k^*(\boldsymbol{\mu}, \boldsymbol{\Lambda})$ $\mathrm{ED}_k(\boldsymbol{\mu}, \boldsymbol{\Sigma})$	Jørgensen,[2] Ch. 3 Jørgensen[1]
Extreme	$\mathrm{XD}^*(\mu, \lambda)$ $\mathrm{XD}(\mu, \sigma^2)$	$\mathrm{XD}_k^*(\boldsymbol{\mu}, \boldsymbol{\Lambda})$ $\mathrm{XD}_k(\boldsymbol{\mu}, \boldsymbol{\Sigma})$	Jørgensen, Goegebeur and Martnez[27] Jørgensen[1]
Geometric	$\mathrm{GD}^*(\mu, \lambda)$ $\mathrm{GD}(\mu, \sigma^2)$	—	Jørgensen and Kokonendji[25]

Acknowledgements

This research was supported by the Danish Natural Science Research Council. We are grateful to Kevin Burke, Gilbert McKenzie, Ingram Olkin and two anonymous referees for useful comments in relation to the paper.

References

1. B. Jørgensen, *Brazilian J. Probab. Statist.*, to appear (2012).
2. B. Jørgensen, *The Theory of Dispersion Models* (Chapman & Hall, London, 1997).
3. B. Jørgensen and S. L. Lauritzen, *J. Multiv. Anal.* **74**, 267 (2000).
4. B. Jørgensen, *J. Roy. Statist. Soc. B* **49**, 127 (1987).
5. N. Balakrishnan and C. D. Lai, *Continuous Bivariate Distributions: Theory and Methods*, second edn. (Springer-Verlag, New York, 2009).
6. S. K. Bar-Lev, D. Bshouty, P. Enis, G. Letac, I.-L. Lu and D. Richards, *J. Theor. Probab* **7**, 883 (1994).
7. P. Holgate, *Biometrika* **51**, 241 (1964).
8. H. Teicher, *Skand. Aktuar.* **37**, 1 (1954).
9. M. Dwass and H. Teicher, *Ann. Math. Statist.* **28**, 461 (1957).
10. A. S. Krishnamoorthy, *Sankyā* **11**, 117 (1951).
11. A. W. Marshall and I. Olkin, *J. Amer. Statist. Assoc.* **80**, 332 (1985).
12. M. A. Hamdan and D. R. Jensen, *Austral. J. Statist.* **18**, 163 (1976).
13. A. Biswas and J.-S. Hwang, *Statist. Probab. Lett.* **60**, 231 (2002).
14. C. B. Edwards and J. Gurland, *J. Amer. Statist. Assoc.* **56**, 503 (1961).
15. K. Subrahmaniam and K. Subrahmaniam, *J. Roy. Statist. Soc. B* **35**, 131 (1973).
16. D. C. Doss, *J. Multiv. Anal.* **9**, 460 (1979).
17. A. M. Mathai and P. G. Moschopoulos, *J. Multiv. Anal.* **39**, 135 (1991).
18. A. Prékopa and T. Szántai, *Water Resources Res.* **14**, 19 (1978).
19. B. Jørgensen, J. R. Martnez and M. Tsao, *Scand. J. Statist.* **21**, 223 (1994).
20. B. Jørgensen, J. R. Martnez and C. G. B. Demtrio, *Lith. Math. J.* **51**, 342 (2011).
21. B. Jørgensen, J. R. Martnez and V. Vinogradov, *Lith. Math. J.* **49**, 399 (2009).

22. E. Furman and Z. Landsman, *Insurance: Mathematics and Economics* **46**, 351 (2010).
23. B. Grigelionis, *Lithuanian Math. J.* **51**, 194 (2011).
24. J. A. Nelder and R. W. M. Wedderburn, *J. Roy. Statist. Soc. A* **135**, 370 (1972).
25. B. Jørgensen and C. C. Kokonendji, *Brazilian J. Probab. Statist.* **25**, 263 (2011).
26. B. Jørgensen, http://imada.sdu.dk/~bentj/ (2011).
27. B. Jørgensen, Y. Goegebeur and J. R. Martnez, *Extremes* **13**, 399 (2010).

STATISTICAL INFERENCE WITH THE LIMITED EXPECTED VALUE FUNCTION

M. KÄÄRIK* and H. KADARIK

ØInstitute of Mathematical Statistics, University of Tartu,
Tartu, Estonia
**E-mail: meelis.kaarik@ut.ee*

One of the common problems in insurance mathematics is that we usually do not see the actual loss variable but certain truncated version of it: the claims payments are limited by sum insured, reinsurance treaties limit the actual claim size for initial insurer, also (fixed amount) deductibles set limits for policy holders, etc. In all these cases a function called *limited expected value function* (or *LEV-function*), defined by $E[X;x] := E(\min(X,x))$, where X is a random variable (claim size), plays an important role. There are many well-known characteristics in insurance that are calculated using this function, which motivated us to study this topic more closely. We reveal some essential properties of this function and describe some important practical applications where it is used. In insurance mathematics, the usage of limited expected value function in premium calculation models is well-known, but in present article we also introduce the method of limited expected value function for measuring the goodness of fit between empirical and theoretical distributions, following the idea proposed by Hogg and Klugman (1984). This is one of the many uses of the limited expected value function and it suits particularly well to the insurance data as it can take into account the censoring of data (if necessary). Also, this method can be used as an alternative or additional tool in case the data is complex and other goodness of fit tests do not give reliable results. The main disadvantage of this method is that the behavior of corresponding test statistic is not thoroughly studied, there are no certain criteria to tell us when the value of this statistic is good enough to say that a proposed distribution fits empirical data well. This problem is of our special interest, several simulations with different distributions are carried out to find the reference values.

Keywords: limited expected value function, distribution fitting, goodness of fit tests, insurance mathematics

1. Limited expected value function

We will first introduce the limited expected value function and its known base properties. We also briefly cover some problems of premium calculation, where the use of limited expected value function is most known, and provide a simple generalization to include all important special cases.

1.1. *Definiton and properties*

Definition 1.1 (Limited expected value function). *For any non-negative random variable X (or corresponding distribution $F(x)$) the limited expected value function $E[X;x]$ is defined by*

$$E[X;x] = E(\min(X,x)) = \int_0^x y dF(y) + x(1 - F(x)), x > 0. \qquad (1)$$

As X is a non-negative random variable, the following alternative formula can also be used:

$$E[X;x] = \int_0^x [1 - F(y)]dy. \qquad (2)$$

Similarly to the formulas above, the empirical limited expected value function can be expressed as

$$E_n[X_n;x] = \frac{1}{n}(\sum_{x_j<x} x_j + \sum_{x_j \geq x} x), \qquad (3)$$

where X_n is the empirical distribution corresponding to sample $(x_1, \ldots, x_n)$.

In the following we assume (although it is not necessary for all properties) that X is continuous, nonnegative and $EX < \infty$. This motivated by the fact that the main interpretation of X is the claim size.

Common general properties with straightforward proofs are:
(1) $E[X;x]$ is continuous, concave and nondecreasing function;
(2) $E[X;x] \to E(X)$, if $x \to \infty$;
(3) $F(x) = 1 - (E[X;x])'$.

For more details see, e.g., Ref. 1 or 2.

It is important to notice that property 3 implies that the LEV-function $E[X;x]$ determines the distribution of X uniquely. It is also easy to prove that for any linear combination $aX + b$ (with $a, b \in \mathcal{R}$ and $a \neq 0$), the corresponding LEV-function is given by

$$E[aX + b; x] = aE\left[X; \frac{x-b}{a}\right] + b.$$

This result is particularly useful in premium calculation, see next subsection for details.

1.2. *Premium calculation*

In insurance mathematics, one usually relates the LEV-function to premium calculation models, where the effect of inflation, deductibles and upper limits of indemnities is of interest. It is particularly important in finding or estimating the expected claim amount, which has central role in most premium calculation problems. This topic is thoroughly covered in Ref. 3 and 4, therefore we just generalize the well-known formulas and introduce one compact formula which includes most of the cases.

Let us have a claim amount random variable X and consider a compensation rule with (fixed amount) deductible d and upper limit u. Also assume that the inflation rate for the period of interest is r. Then the following random variable $X_{d,u,r}$ describing the compensation amount can be defined:

$$X_{d,u,r} = \begin{cases} 0, & \text{if } (1+r)X < d, \\ (1+r)X - d, & \text{if } d \leq (1+r)X \leq u, \\ u - d, & \text{if } (1+r)X > u. \end{cases}$$

In that case the expected claim amount (in case claim occurs) can be calculated as follows:

$$E(X_{d,u,r}) = (1+r)\left(E\left[X;\frac{u}{1+r}\right] - E\left[X;\frac{d}{1+r}\right]\right). \tag{4}$$

The formula (4) clearly stresses the importance of the limited expected value function in premium calculation. Notice also that setting $r = 0$, $d = 0$ or $u \to \infty$ lead to well-known special cases.

It is also worth mentioning that the LEV-function is also closely related to integrated tail distribution, which is a useful tool from extreme value theory and applicable to answer questions related to ruin theory. Integrated tail distribution is used in renewal risk models to describe ruin probabilities, in queuing theory as limiting distribution for waiting times and busy periods, also in problems related to random walks, and more.[5–7]

2. Distribution fitting

2.1. *Setup*

Consider the following situation: let us have an (ordered) sample $(x_1, \ldots, x_n)$ with c censored observations and let us have some proposed theoretical distribution F to be fitted to that data. An idea to measure the goodness of fit of distribution F and the empirical distribution based on data is introduced in Ref. 3:

- Calculate the values of empirical LEV-function $E_n[X; x_i]$ by (3) and theoretical LEV-function $E[X; x_i]$ by (1) or (2) at each uncensored sample point , $i = 1, ..., n-c$.
- Find the (relative) differences d_i at each uncensored sample point and form a difference vector $\vec{D}$ as follows:

$$d_i = \frac{E[X, x_i] - E_n[X_n; x_i]}{E[X; x_i]}, \; i = 1, ..., n-c,$$
$$\vec{D} = (d_1, ..., d_{n-c}).$$

Obviously, the smaller the difference vector $\vec{D}$ (the closer to 0-vector) the better fit we have. Natural question to investigate is when is the vector $\vec{D}$ small enough to say that the corresponding distribution fits well to data.

To answer this question we propose the following simple error statistics

$$e_1 = \frac{1}{n-c} \sum_{i=1}^{n-c} |d_i|, \tag{5}$$

$$e_2 = \sqrt{\frac{1}{n-c} \sum_{i=1}^{n-c} d_i^2}. \tag{6}$$

One can think of e_1 and e_2 as realizations from random variables $\mathcal{E}_1$ and $\mathcal{E}_2$ and try to study their behaviour. We are mostly interested in finding estimates for expected values and critical values. The problem of the estimation of p-values while testing goodness of fit with given data will also be addressed. To find these estimates we will construct two parametric bootstrap algorithms.

First, in order to find critical values for a fixed theoretical distribution we can use the following algorithm.

Algorithm 2.1. ***Parametric bootstrap for critical values.***

(1) Fix the theoretical distribution F and its parameter(s) θ.
(2) Fix the number of iterations k and the size of sample to be simulated n.
(3) Repeat the following steps k times ($j = 1, \ldots, k$):

(a) simulate sample of size n from $F(\cdot|\theta)$;
(b) calculate $e_1^{(j)}$ and $e_2^{(j)}$ from the sample using formulas (5) and (6).

(4) *Find the mean values $\bar{e}_1$ and $\bar{e}_2$ and required critical values $e_{1,\alpha}$ and $e_{2,\alpha}$ from the empirical distributions of $(e_1^{(1)}, \ldots, e_1^{(k)})$ and $(e_2^{(1)}, \ldots, e_2^{(k)})$.*

As a result of algorithm 2.1 we get 2 samples of size k: $(e_1^{(1)}, \ldots, e_1^{(k)})$ and $(e_2^{(1)}, \ldots, e_2^{(k)})$ that can be considered as realizations from theoretical random variables $\mathcal{E}_1$ and $\mathcal{E}_2$, and the sample characteristics can be considered as estimates for characteristics of $\mathcal{E}_1$ and $\mathcal{E}_2$:

- the mean values $\bar{e}_1$ and $\bar{e}_2$ are estimates for expectations $E(\mathcal{E}_1)$ and $E(\mathcal{E}_2)$;
- the critical values $e_{1,\alpha}$ and $e_{2,\alpha}$ are estimates for probabilities $\mathbf{P}\{\mathcal{E}_1 > \alpha\}$ and $\mathbf{P}\{\mathcal{E}_2 > \alpha\}$.

In case we need to test the goodness of fit of proposed theoretical distribution with given data, we can construct similar parametric bootstrap algorithm to estimate the p-values.

Algorithm 2.2. ***Parametric bootstrap for estimation of p-value.***

(1) *Estimate the parameter(s) θ of the distribution F to be fitted to data.*
(2) *Calculate e_1^* and e_2^* by formulas (5) and (6) using empirical distribution from data and $F(\cdot|\theta)$ as theoretical distribution.*
(3) *Fix the number of iterations k and repeat the following steps k times $(j = 1, \ldots, k)$:*
    (a) *simulate sample of size n (usually equal to original sample size) from $F(\cdot|\theta)$;*
    (b) *calculate $e_1^{(j)}$ and $e_2^{(j)}$ from the simulated sample, using $F(\cdot|\theta)$ as the theoretical distribution.*
(4) *The estimates for p-values are given by the following formulas: $p_1 = \frac{1}{k}\sum_{j=1}^{k} I_{\{e_1^{(j)} > e_1^*\}}$ and $p_2 = \frac{1}{k}\sum_{j=1}^{k} I_{\{e_2^{(j)} > e_2^*\}}$.*

Here I represents an indicator function (i.e. its value is 1 if corresponding condition is fulfilled and 0 otherwise). The algorithm 2.2 allows us to test the following pair of hypothesis:

$$\begin{cases} H_0: & \text{data follows the theoretical distribution } F(\cdot|\theta), \\ H_1: & \text{data does not follow the theoretical distribution } F(\cdot|\theta). \end{cases}$$

The null hypothesis will be rejected with probability $1 - p_1$ or $1 - p_2$ (depending on which criterion we use).

It is also possible to construct an algorithm for estimation of test power similarly to algorithms 2.1 and 2.2:

Algorithm 2.3. ***Parametric bootstrap for power estimation.***

(1) Choose a theoretical distribution F_{H_0} and an alternative distribution F_{H_1}.
(2) Fix the number of iterations k and the sample size n.
(3) Estimate the critical values $e_{1,\alpha}$ and $e_{2,\alpha}$ for F_{H_0} (e.g. by using algorithm 2.1).
(4) Repeat the following steps k times ($j = 1, \ldots, k$):
   (a) simulate sample of size n from F_{H_1};
   (b) find $e_1^{(j)}$ and $e_2^{(j)}$ from the simulated sample, using F_{H_0} as the theoretical distribution.
(5) The rejection proportions are given by $r_{1,\alpha} = \frac{1}{k}\sum_{j=1}^{k} I_{\{e_1^{(j)} > e_{1,\alpha}\}}$ and $r_{2,\alpha} = \frac{1}{k}\sum_{j=1}^{k} I_{\{e_2^{(j)} > e_{2,\alpha}\}}$.

At each iteration step $j = 1 \ldots k$ in algorithm 2.3 step $4a$, the pair of hypotheses corresponding to this schema is

$$\begin{cases} H_0: & \text{generated sample follows the distribution } F_{H_0}, \\ H_1: & \text{generated sample does not follow the distribution } F_{H_0}. \end{cases}$$

By constrution the sample comes from alternative distribution F_{H_1} instead of F_{H_0} and the rejection proportions $r_{1,\alpha}$ and $r_{2,\alpha}$ describe how often we actually conclude that.

2.2. *Simulations*

A simulation study was carried out to find the reference values for most common distributions used in insurance practice and also to study the influence of distribution parameters and generated sample size n to error characteristics e_1 and e_2 (or, equivalently, to corresponding random variables $\mathcal{E}_1$ and $\mathcal{E}_2$). Lognormal, Gamma, Weibull and Pareto distributions were included to study, for brevity only the results with lognormal distributions are given in more detail, an overview of the results with other distributions can be found in appendix.

In case of lognormal distribution ($X \sim LnN(\mu, \sigma)$), the expression for LEV-function is

$$E[X; x] = e^{\mu + \frac{\sigma^2}{2}} \Phi\left(\frac{\ln x - \mu - \sigma^2}{\sigma}\right) + x\left\{1 - \Phi\left(\frac{\ln x - \mu}{\sigma}\right)\right\}.$$

We follow the steps described in algorithm 2.1, and apply the procedure for 4 different sample sizes n (100, 1000, 10 000 and 1 000 000) and 3 different shape parameters σ $(0.1, 1, 10)$. The the size of bootstrap sample (i.e. the number of iterations) k is usually 10 000, for some cases also $k = 1\ 000\ 000$ is used. As the scale parameter μ did not affect the results (tested with 7 different values of μ), we can take arbitrary μ, for simplicity we can assume that $\mu = 0$.

Table 1. Results for e_1, simulations from $LnN(0, \sigma)$, no censoring.

σ	n	$\bar{e}_1$	$e_{1,0.1}$	$e_{1,0.05}$	$e_{1,0.01}$
0.1	100	0.0044	0.0085	0.010	0.014
	1000	0.0014	0.0027	0.0032	0.0042
	10000	0.00043	0.00084	0.0010	0.0013
	1000000	0.000043	0.000084	0.00010	0.00013
1	100	0.035	0.064	0.077	0.11
	1000	0.011	0.020	0.024	0.031
	10000	0.0034	0.0063	0.0076	0.010
	1000000	0.00034	0.00064	0.00076	0.0010
10	100	0.18	0.24	0.33	0.82
	1000	0.040	0.059	0.070	0.11
	10000	0.011	0.017	0.020	0.026
	1000000	0.0011	0.0016	0.0019	0.0024

Table 2. Results for e_2, simulations from $LnN(0, \sigma)$, no censoring.

σ	n	$\bar{e}_2$	$e_{2,0.1}$	$e_{2,0.05}$	$e_{2,0.01}$
0.1	100	0.0051	0.0098	0.012	0.016
	1000	0.0016	0.0031	0.0037	0.0048
	10000	0.00051	0.00098	0.0011	0.0015
	1000000	0.000051	0.00010	0.00011	0.00015
1	100	0.044	0.080	0.095	0.13
	1000	0.014	0.025	0.029	0.038
	10000	0.0043	0.0078	0.0092	0.012
	1000000	0.00043	0.00078	0.00092	0.0012
10	100	0.81	0.74	1.38	6.2
	1000	0.20	0.23	0.40	1.8
	10000	0.082	0.065	0.12	0.54
	1000000	0.0061	0.0057	0.0098	0.042

As seen from tables 1 and 2, increasing the sample size m times decreases the estimates for mean values and critical values approximately $\sqrt{m}$ times. Also, the smaller variance, the stronger the above-mentioned tendency holds. Same is true for all other distributions used (see Appendix A for details).

In the following we also present the simulation results with censored samples, where the censoring rules are constructed so that a sample is generated from $LnN(\mu, 1)$ and either 0.9-, 0.75-, or 0.5-quantile of the distribution is chosen for censoring threshold, thus resulting the average censoring proportions $c = 0.1n$, $0.25n$ or $0.5n$. Similarly to non-censored case, the scale parameter μ does not affect the results, so we can simply use the $LnN(0, 1)$ distribution for simulations.

Table 3. Results for e_1, simulations from $LnN(0, 1)$, with censoring.

c	n	$\bar{e}_1$	$e_{1,0.1}$	$e_{1,0.05}$	$e_{1,0.01}$
0.10n	100	0.030	0.056	0.067	0.093
	1000	0.0094	0.018	0.021	0.027
	10000	0.0030	0.0057	0.0067	0.0089
	100000	0.00094	0.0018	0.0021	0.0028
0.25n	100	0.025	0.047	0.057	0.077
	1000	0.0079	0.015	0.018	0.023
	10000	0.0025	0.0047	0.0056	0.0075
	100000	0.00079	0.0015	0.0018	0.0023
0.50n	100	0.018	0.035	0.041	0.055
	1000	0.0057	0.011	0.013	0.017
	10000	0.0018	0.0035	0.0041	0.0055
	100000	0.00057	0.0011	0.0013	0.0017

Table 4. Results for e_2, simulations from $LnN(0, 1)$, with censoring.

c	n	$\bar{e}_2$	$e_{2,0.1}$	$e_{2,0.05}$	$e_{2,0.01}$
0.10n	100	0.036	0.066	0.079	0.11
	1000	0.011	0.021	0.025	0.032
	10000	0.0036	0.0067	0.0080	0.010
	100000	0.0011	0.0021	0.0025	0.0032
0.25n	100	0.029	0.055	0.065	0.086
	1000	0.0093	0.018	0.021	0.027
	10000	0.0029	0.0054	0.0065	0.0086
	100000	0.00093	0.0018	0.0021	0.0027
0.50n	100	0.020	0.039	0.047	0.062
	1000	0.0066	0.012	0.015	0.020
	10000	0.0021	0.0040	0.0047	0.0062
	100000	0.00065	0.0013	0.0015	0.0019

The simulations with censoring (tables 3 and 4) point out the same tendencies as before: the scale parameter does not affect the estimates of mean values and critical values and increasing the sample size m times decreases the estimates for mean values and critical values approximately $\sqrt{m}$ times. Moreover, the fit with cases with larger variances is even better than for non-censored case, which may be explained by the fact that by censoring some variability is removed.

2.3. *Conclusions*

In conclusion, the behaviour of test statistics was similar for all the distributions used in simulations (lognormal, Gamma, Weibull and Pareto). The following general tendencies can be pointed out:

- scale parameter does not affect the estimates for mean values of test statistics $\bar{e}_1$ and $\bar{e}_2$ and critical values of test statistics $e_{1,\alpha}$ and $e_{2,\alpha}$ at any chosen significance level α $(0.1, 0.05$ and $0.1)$.
- the size of sample n has the following effect to test statistics: mean values $\bar{e}_1\sqrt{n}$ (and $\bar{e}_2\sqrt{n}$) are are approximately equal for different sample sizes n (especially when theoretical variance is small), same holds for critical values $e_{1,\alpha}$ and $e_{2,\alpha}$ at significance levels $\alpha = 0.1, 0.05, 0.01$.
- similar results hold for censored cases, also, the dependence from sample size n is even more clear in cases with larger variance.

Acknowledgments

This research is supported by Estonian Science Foundation Grants No 7313 and No 8802 and by Targeted Financing Project SF0180015s12. The authors also thank the two referees for their helpful comments and suggestions.

Appendix A. Reference values for Gamma, Weibull and Pareto distributions

In this section we briefly list the simulation results obtained for Gamma, Weibull and Pareto distributions. The formulas of corresponding distribution functions and LEV-functions are given to prevent the ambiguities of different parametrisations. The formulas of LEV-functions for different distributions can be found in various textbooks, proofs are straightforward.[1,2]

Appendix A.1. *Gamma distribution*

Let us have a Gamma-distributed random variable $X \sim \Gamma(\alpha, \beta)$, i.e. its distribution function is $F(x|\alpha, \beta) = \int_0^x \beta^\alpha y^{\alpha-1} \frac{e^{-\beta y}}{\Gamma(\alpha)} dy$, where $\alpha > 0$ and

$\beta > 0$. Then the corresponding LEV-function has the following form:

$$E[X;x] = \frac{\alpha}{\beta} F(x|\alpha+1,\beta) + x(1 - F(x|\alpha,\beta)).$$

The reference values for constructed test-statistics e_1 and e_2 (see (5) and (6)) for a Gamma-distributed random variable are found in tables A1 and A2, the results are obtained using algorithm 2.1 with the number of iterations $k = 10\ 000$. As the scale parameter β did not affect the result, it is taken $\beta = 1$ for simplicity.

Table A1. Results for e_1, simulations from $\Gamma(\alpha, 1)$, no censoring.

α	n	$\bar{e}_1$	$e_{1,0.1}$	$e_{1,0.05}$	$e_{1,0.01}$
	100	0.039	0.074	0.089	0.12
1	1000	0.012	0.023	0.027	0.036
	10000	0.0039	0.0073	0.0087	0.0011
	100	0.019	0.037	0.044	0.058
5	1000	0.0061	0.0011	0.014	0.018
	10000	0.0019	0.0037	0.0044	0.0058
	100	0.014	0.027	0.032	0.042
10	1000	0.0043	0.0083	0.010	0.013
	10000	0.0014	0.0027	0.0032	0.0042

Table A2. Results for e_2, simulations from $\Gamma(\alpha, 1)$, no censoring.

α	n	$\bar{e}_2$	$e_{2,0.1}$	$e_{2,0.05}$	$e_{2,0.01}$
	100	0.046	0.086	0.10	0.14
1	1000	0.015	0.0027	0.032	0.042
	10000	0.0046	0.0085	0.010	0.013
	100	0.022	0.042	0.050	0.067
5	1000	0.0071	0.013	0.016	0.021
	10000	0.0022	0.0042	0.0050	0.0066
	100	0.016	0.030	0.036	0.048
10	1000	0.0050	0.0095	0.011	0.015
	10000	0.0016	0.0030	0.0036	0.0047

Appendix A.2. *Weibull distribution*

Let us have a Weibull-distributed random variable $X \sim Weibull(\alpha, \beta)$, i.e. its distribution function is $F(x|\alpha,\beta) = 1 - e^{-(\beta x)^{\alpha}}$, where $\alpha > 0$ and $\beta > 0$.

Then the corresponding LEV-function has the following form:

$$E[X;x] = \frac{\Gamma(1+\frac{1}{\alpha})}{\beta}\Gamma\left((\beta x)^{\alpha}, 1+\frac{1}{\alpha}\right) + xe^{-(\beta x)^{\alpha}},$$

where $\Gamma(\alpha)$ is gamma-function and $\Gamma(x,\alpha)$ is incomplete gamma-function:

$$\Gamma(x,\alpha) = \frac{1}{\Gamma(\alpha)}\int_0^x y^{\alpha-1}e^{-y}dy.$$

The reference values for constructed test-statistics e_1 and e_2 for a Weibull-distributed random variable are found in tables A3 and A4, the results are obtained using algorithm 2.1, the number of iterations k is 10 000. As the scale parameter β did not affect the result, it is taken $\beta = 1$ for simplicity.

Table A3. Results for e_1, simulations from $Weibull(\alpha, 1)$, no censoring.

α	n	$\bar{e}_1$	$e_{1,0.1}$	$e_{1,0.05}$	$e_{1,0.01}$
	100	0.039	0.073	0.088	0.12
1	1000	0.012	0.023	0.027	0.036
	10000	0.0039	0.0073	0.0087	0.0011
	100	0.012	0.024	0.028	0.037
5	1000	0.038	0.074	0.088	0.012
	10000	0.0012	0.0024	0.0028	0.0037
	100	0.0065	0.013	0.015	0.020
10	1000	0.0021	0.0041	0.0048	0.0063
	10000	0.00065	0.0013	0.0015	0.0020

Table A4. Results for e_2, simulations from $Weibull(\alpha, 1)$, no censoring.

α	n	$\bar{e}_2$	$e_{2,0.1}$	$e_{2,0.05}$	$e_{2,0.01}$
	100	0.047	0.085	0.10	0.13
1	1000	0.015	0.0027	0.032	0.041
	10000	0.0047	0.0085	0.010	0.013
	100	0.013	0.026	0.031	0.040
5	1000	0.0042	0.0081	0.0096	0.013
	10000	0.0013	0.0026	0.0031	0.0040
	100	0.0072	0.014	0.017	0.022
10	1000	0.0023	0.0044	0.0052	0.0068
	10000	0.00072	0.0014	0.0017	0.0021

Appendix A.3. *Pareto distribution*

Let us have a Pareto-distributed random variable $X \sim Pa(\alpha, \lambda)$, i.e. its distribution function is $F(x|\lambda, \alpha) = 1 - \left(\frac{\lambda}{x+\lambda}\right)^{\alpha}$, where $\alpha > 0$ and $\lambda > 0$. Then the corresponding LEV-function has the following form:

$$E[X; x] = \frac{\lambda - \lambda^{\alpha}(\lambda + x)^{1-\alpha}}{\alpha - 1}.$$

The reference values for constructed test-statistics e_1 and e_2 for a Pareto-distributed random variable are found in tables A5 and A6. The results are obtained using algorithm 2.1, where the number of iterations k is 10 000. As the scale parameter λ did not affect the results, it can be taken arbitrary, $\lambda = 1$ is chosen for simplicity.

Table A5. Results for e_1, simulations from $Pa(\alpha, 1)$, no censoring.

α	n	$\bar{e}_1$	$e_{1,0.1}$	$e_{1,0.05}$	$e_{1,0.01}$
	100	0.041	0.076	0.092	0.13
5	1000	0.013	0.024	0.028	0.037
	10000	0.0041	0.0076	0.0091	0.0012
	100	0.040	0.075	0.090	0.12
10	1000	0.013	0.024	0.028	0.037
	10000	0.0040	0.0073	0.0088	0.011

Table A6. Results for e_2, simulations from $Pa(\alpha, 1)$, no censoring.

α	n	$\bar{e}_2$	$e_{2,0.1}$	$e_{2,0.05}$	$e_{2,0.01}$
	100	0.050	0.091	0.11	0.15
5	1000	0.016	0.029	0.034	0.044
	10000	0.0050	0.0091	0.011	0.014
	100	0.049	0.088	0.11	0.14
10	1000	0.015	0.028	0.033	0.042
	10000	0.0048	0.0088	0.010	0.013

References

1. P. Čižek, W. Härdle and R. Weron (2005), *Statistical tools for finance and insurance*. Springer, Berlin.
2. C. Chen, W. Härdle and A. Unwin (2008), *Handbook Of Data Visualization*, Springer, Berlin.
3. R. V. Hogg and S. A. Klugman (1984), *Loss distributions*. Wiley, New York.
4. S. A. Klugman, H. H. Panjer and G. E. Willmot (2004), *Loss Models: From Data to Decisions, 2nd Ed.* Wiley, New York.
5. S. Asmussen (2000), *Ruin Probabilities*. World Scientific, Singapore.
6. S. Asmussen (2003), *Applied probability and queues*. Springer, New York.
7. P. Embrechts, T. Mikosch and C. Klüppelberg (1997), *Modelling Extremal Events: for Insurance and Finance*. Springer, Berlin.

SHRINKAGE ESTIMATION VIA PENALIZED LEAST SQUARES IN LINEAR REGRESSION WITH AN APPLICATION TO HIP FRACTURE TREATMENT COSTS

A. LISKI

Department of Signal Processing, Tampere University of Technology, Tampere, Finland
E-mail: antti.liski@tut.fi

E. P. LISKI

Department of Mathematics and Statistics, University of Tampere, Tampere, Finland
E-mail: erkki.liski@uta.fi

U. HÄKKINEN

Centre for Health and Social Economics, National Institute for Health and Welfare, Helsinki, Finland
E-mail: unto.hakkinen@thl.fi

In this paper, we consider the problem of averaging across least squares estimates obtained from a set of models. Existing model averaging (MA) methods usually require estimation of a single weight for each candidate model. However, in applications the number of candidate models may be huge. Then the approach based on estimation of single weights becomes computationally infeasible. Utilizing a connection between shrinkage estimation and model weighting we present an accurate and computationally efficient MA estimation method. The performance of our estimators is displayed in simulation experiments which utilize a realistic set up based on real data.

Keywords: Model averaging, Model selection, Mean square error, Efficiency bound, Simulation experiment

1. Introduction

Our framework is the linear model

$$\boldsymbol{y} = \mathbf{X}\boldsymbol{\beta} + \mathbf{Z}\boldsymbol{\gamma} + \boldsymbol{\varepsilon}, \qquad \boldsymbol{\varepsilon} \sim (\mathbf{0}, \sigma^2\mathbf{I}_n), \tag{1}$$

where $\mathbf{X}$ and $\mathbf{Z}$ are $n \times p$ and $n \times m$ matrices of nonrandom regressors, $(\mathbf{X}, \mathbf{Z})$ is assumed to be of full column-rank $p + m < n$, $\boldsymbol{\beta}$ and $\boldsymbol{\gamma}$ are $p \times 1$

and $m \times 1$ vectors of unknown parameters. Our interest is in the effect of $\mathbf{X}$ on $\boldsymbol{y}$, that is, we want to estimate $\boldsymbol{\beta}$ while the role of $\mathbf{Z}$ is to improve the estimation of $\boldsymbol{\beta}$. It is known that dropping z-variables from the model decreases the variance of the least squares (LS) estimator of the β-parameters. However, after elimination of variables, the estimates are biased if the full model is correct. In certain applications the model (1) can also be interpreted as an analysis of covariance (ANCOVA) model which is a technique that sits between the analysis of variance and regression analysis. However, ANCOVA is only a special instance of the general regression model (1).

We introduce a set of shrinkage estimators for the regression coefficients $\boldsymbol{\beta}$ in the class of penalized least squares estimators. The efficiency bound of estimators with respect to the mean square (MSE) error criterion within this shrinkage class is known. We search for the estimators whose MSE is uniformly close to the efficiency bound. It turns out that many interesting known estimators belong to this class, for example the soft thresholding and the firm thresholding estimators, non-negative garrote, LASSO and SCAD estimators. On the other hand, for example the hard thresholding rule (pre testing) and the ridge estimator do not belong to this shrinkage class. In Section 2 we present the canonical form of the model (1). The problem of model selection and averaging is introduced in Section 3. We characterize our class of shrinkage estimators within the set of penalized least squares estimators in Subsection 4.1, and the main result on shrinkage and penalised least squared estimation is given in Subsection 4.2. Examples of good alternative penalized least squares estimators, which are also shrinkage estimators, are introduced in Subsection 4.3. A real data application is given in Section 5 and the results of the simulation experiments are reported in Section 6.

2. The model

We will work with the canonical form of the model (1) where z-variables are orthogonalized by writing the systematic part of the model (1) as

$$\begin{aligned}\mathbf{X}\boldsymbol{\beta} + \mathbf{Z}\boldsymbol{\gamma} &= \mathbf{X}\boldsymbol{\beta} + \mathbf{P}\mathbf{Z}\boldsymbol{\gamma} + (\mathbf{I} - \mathbf{P})\mathbf{Z}\boldsymbol{\gamma} \\ &= \mathbf{X}\boldsymbol{\alpha} + \mathbf{M}\mathbf{Z}\boldsymbol{\gamma},\end{aligned} \tag{2}$$

where

$$\mathbf{P} = \mathbf{X}(\mathbf{X}'\mathbf{X})^{-1}\mathbf{X}' \quad \text{and} \quad \mathbf{M} = \mathbf{I}_n - \mathbf{P} \tag{3}$$

are symmetric idempotent matrices and $\boldsymbol{\alpha} = \boldsymbol{\beta} + (\mathbf{X}'\mathbf{X})^{-1}\mathbf{X}'\mathbf{Z}\boldsymbol{\gamma}$. Since $(\mathbf{MZ})'\mathbf{MZ} = \mathbf{Z}'\mathbf{MZ}$ is positive definite,[1] then there exists a nonsingular matrix $\mathbf{C}$ such that[2]

$$\mathbf{C}'\mathbf{Z}'\mathbf{M}\mathbf{Z}\mathbf{C} = (\mathbf{M}\mathbf{Z}\mathbf{C})'(\mathbf{M}\mathbf{Z}\mathbf{C}) = \mathbf{U}'\mathbf{U} = \mathbf{I}_m. \tag{4}$$

In (4) $\mathbf{U} = \mathbf{M}\mathbf{Z}\mathbf{C}$ denotes the matrix of orthogonal canonical auxiliary regressors. Introducing the canonical auxiliary parameters $\boldsymbol{\theta} = \mathbf{C}^{-1}\boldsymbol{\gamma}$ we can write in (2)

$$\mathbf{M}\mathbf{Z}\boldsymbol{\gamma} = \mathbf{M}\mathbf{Z}\mathbf{C}\mathbf{C}^{-1}\boldsymbol{\gamma} = \mathbf{U}\boldsymbol{\theta}.$$

There are advantages working with $\boldsymbol{\theta}$ instead of $\boldsymbol{\gamma}$, and we can always get $\boldsymbol{\gamma}$ from $\boldsymbol{\gamma} = \mathbf{C}\boldsymbol{\theta}$.

We are interested in estimation of $\boldsymbol{\beta}$, whereas $\boldsymbol{\theta}$ contains auxiliary parameters. Let $\mathcal{M}_0$ denote the fully restricted model without any auxiliary regressors and $\mathcal{M}_M$ the unrestricted model containing all auxiliary regressors as follows

$$\mathcal{M}_0 : \{\boldsymbol{y}, \mathbf{X}\boldsymbol{\beta}, \sigma^2\mathbf{I}_n\} \text{ and } \mathcal{M}_M : \{\boldsymbol{y}, \mathbf{X}\boldsymbol{\alpha} + \mathbf{U}\boldsymbol{\theta}, \sigma^2\mathbf{I}_n\}. \tag{5}$$

Now in θ-parametrization we write $\boldsymbol{\alpha} = \boldsymbol{\beta} + (\mathbf{X}'\mathbf{X})^{-1}\mathbf{X}'\mathbf{Z}\mathbf{C}\mathbf{C}^{-1}\boldsymbol{\gamma} = \boldsymbol{\beta} + \mathbf{Q}\boldsymbol{\theta}$, where $\mathbf{Q} = (\mathbf{X}'\mathbf{X})^{-1}\mathbf{X}'\mathbf{Z}\mathbf{C}$. The model $\mathcal{M}_M$ is orthogonal such that $\mathbf{X}'\mathbf{U} = \mathbf{0}$ and $(\mathbf{X}, \mathbf{U})$ is of full column rank. Then the least squares (LS) estimators of $\boldsymbol{\alpha}$ and $\boldsymbol{\theta}$ from the model $\mathcal{M}_M$ are

$$\begin{aligned}\hat{\boldsymbol{\alpha}} &= (\mathbf{X}'\mathbf{X})^{-1}\mathbf{X}'\boldsymbol{y},\\ \hat{\boldsymbol{\theta}} &= \mathbf{U}'\boldsymbol{y}.\end{aligned}$$

Let $\hat{\boldsymbol{\beta}}_0$ denote the LS estimator of $\boldsymbol{\beta}$ under the restricted model $\mathcal{M}_0$ and note that $\hat{\boldsymbol{\alpha}} \equiv \hat{\boldsymbol{\beta}}_0$. The correspondence between the vectors $(\boldsymbol{\alpha}', \boldsymbol{\theta}')$ and $(\boldsymbol{\beta}', \boldsymbol{\theta}')$ is one-to-one, and consequently the same correspondence holds between their LS estimates. Hence the LS estimate of $\boldsymbol{\beta}$ under the unrestricted model $\mathcal{M}_M$ is[1,3]

$$\begin{aligned}\hat{\boldsymbol{\beta}}_M &= \hat{\boldsymbol{\alpha}} - \mathbf{Q}\hat{\boldsymbol{\theta}}\\ &= \hat{\boldsymbol{\beta}}_0 - \mathbf{Q}\hat{\boldsymbol{\theta}}.\end{aligned}$$

In the unrestricted model $\mathcal{M}_M$ in (5) there are m components of $\boldsymbol{\theta}$, and 2^m submodels are obtained by setting various subsets of the elements $\theta_1, \ldots, \theta_m$ of $\boldsymbol{\theta}$ equal to zero. These 2^m models $\mathcal{M}_0, \ldots, \mathcal{M}_M$ can be written as

$$\mathcal{M}_i : \{\boldsymbol{y}, \mathbf{X}\boldsymbol{\alpha} + \mathbf{U}_i\boldsymbol{\theta}, \sigma^2\mathbf{I}_n\},$$

where $\mathbf{U}_i = \mathbf{U}\mathbf{W}_i$ and $\mathbf{W}_i = \operatorname{diag}(w_{i1}, \ldots, w_{im})$, $i = 0, 1, \ldots, M$ are $m \times m$ diagonal matrices with diagonal elements $w_{ij} \in \{0, 1\}, j = 1, \ldots, m$, and $M = 2^m - 1$. We may suppose that the models are in increasing order with respect to diagonal elements of $\mathbf{W}_i$ when the diagonals are interpreted as m-digit binary numbers $w_{i1} \dots w_{im}$, $i = 0, 1, \dots, M$. Then the indices $1, \dots, M$ are associated with the diagonals as follows

$$\begin{aligned} &0 \to 00\dots0, \quad 1 \to 0\dots01, \quad 2 \to 0\dots010, \quad 3 \to 0\dots011, \dots, \\ &M-1 \to 11\dots10, \quad M \to 11\dots11, \end{aligned} \tag{6}$$

where the number of models is $M + 1 = 2^m$. Standard theory of LS estimation with linear restrictions[1,3] yields the restricted LS estimators

$$\hat{\boldsymbol{\beta}}_i = \hat{\boldsymbol{\beta}}_0 - \mathbf{Q}\mathbf{W}_i\hat{\boldsymbol{\theta}} \tag{7}$$

for $\boldsymbol{\beta}$ under the models $\mathcal{M}_i, 0 \le i \le M$.

3. Model selection and averaging

The aim of model selection (MS) is to choose a model $\mathcal{M}_i, 0 \le i \le M$, from the set of candidate models $\mathcal{M}_0, \dots, \mathcal{M}_M$. Given an MS procedure S, the associated post MS estimator may be represented as

$$\hat{\boldsymbol{\beta}}_S = \sum_{i=0}^{M} \mathbb{I}(S = \mathcal{M}_i)\hat{\boldsymbol{\beta}}_i, \tag{8}$$

where $\hat{\boldsymbol{\beta}}_i$ denotes the LS estimator of $\boldsymbol{\beta}$ under $\mathcal{M}_i$ and $\mathbb{I}(\cdot)$ is the indicator function with the value 1 for the selected model and 0 for all other models. Akaike's information criterion AIC[4] and Bayesian information criterion BIC,[5] as well as the minimum description length (MDL) principle,[6–8] for example, are well known MS criteria. However, traditionally by far the most common selection approach in practice is to carry out a sequence of tests in order to identify the nonzero regression coefficients and select the corresponding regressors. Forward selection, backward elimination, and stepwise regression are the best known examples of these techniques.[9] It is not unusual that $\boldsymbol{\beta}$ is estimated from the selected model and the properties of the estimator are reported as if estimation had not been preceeded by model selection. Deleting variables from a model increases bias and decreases variance. To minimize the mean square error (MSE) of estimation, a balance must be attained between the bias due to omitted variables and the variance due to parameter estimation.

Model averaging (MA) offers a more general way of weighting models than just by means of indicator functions like in model selection (8). Let $\boldsymbol{\lambda} = (\lambda_0, \lambda_1, \dots, \lambda_M)'$ be a vector of nonnegative weights which sum to one and thus $\boldsymbol{\lambda}$ lies on the $\mathbb{R}^{M+1}$ unit simplex

$$\Delta^{M+1} = \{\boldsymbol{\lambda} \in [0,1]^{M+1} : \sum_{i=0}^{M} \lambda_i = 1\}. \tag{9}$$

Then a model averaging LS estimator for $\boldsymbol{\beta}$ takes the form

$$\begin{aligned}\tilde{\boldsymbol{\beta}} &= \sum_{i=0}^{M} \lambda_i \hat{\boldsymbol{\beta}}_i = \sum_{i=0}^{M} \lambda_i (\hat{\boldsymbol{\beta}}_0 - \mathbf{Q}\mathbf{W}_i \hat{\boldsymbol{\theta}}) \\ &= \hat{\boldsymbol{\beta}}_0 - \mathbf{Q}\mathbf{W}\hat{\boldsymbol{\theta}},\end{aligned} \tag{10}$$

where $\mathbf{W} = \sum_{i=0}^{M} \lambda_i \mathbf{W}_i$. Hansen[10] shows that a LS model averaging estimator like (10) can achieve lower MSE than any individual estimator (7). Magnus et al.[11] introduced the LS model averaging estimator (10) and called it weighted-average LS (WALS).

Magnus et al.[11] assume that the weights $\lambda_0, \lambda_1, \ldots, \lambda_M$ in (9) are random and they depend on least squares residuals $\mathbf{M}\boldsymbol{y}$, i.e.

$$\lambda_i = \lambda_i(\mathbf{M}\boldsymbol{y}), \quad i = 0, 1, \ldots, M. \tag{11}$$

Note especially that $\hat{\boldsymbol{\theta}}$ is a function of $\mathbf{M}\boldsymbol{y}$. Similarly in model selection (8), $I(S = \mathcal{M}_i) = \lambda_i(\mathbf{M}\boldsymbol{y}) = 1$ for exactly one $i \in \{1, \ldots, M\}$. Thus model selection is a special case of model averaging. Note that the selection matrices $\mathbf{W}_i$, $0 \le i \le M$ are nonrandom $m \times m$ diagonal matrices whereas $\mathbf{W}$ is a random $m \times m$ diagonal matrix with diagonal elements

$$\boldsymbol{w} = (w_1, \ldots, w_m)', \quad 0 \le w_i \le 1,\ i = 1, \ldots, m. \tag{12}$$

For example, when $m = 3$, we have $M+1 = 2^3$ models to compare. If we use the indexing given in (6), the diagonal elements of the selection matrices $\mathbf{W}_i, i = 0, 1, \ldots, 7$ are

$$\begin{array}{cccc} 0:000 & 1:001 & 2:010 & 3:011 \\ 4:100 & 5:101 & 6:110 & 7:111 \end{array}$$

and hence the diagonal entries of $\mathbf{W}$

$$w_1 = \lambda_4 + \lambda_5 + \lambda_6 + \lambda_7, \quad w_2 = \lambda_2 + \lambda_3 + \lambda_6 + \lambda_7, \quad w_3 = \lambda_1 + \lambda_3 + \lambda_5 + \lambda_7.$$

are random variables such that $0 \le w_i \le 1,\ i = 1, 2, 3$.

The equivalence theorem of Danilov and Magnus[12] provides a useful representation for the expectation, variance and MSE of the WALS estimator $\tilde{\boldsymbol{\beta}}$ given in (10). The theorem was proved under the assumptions that the disturbances $\varepsilon_1, \ldots, \varepsilon_n$ are i.i.d. $\mathrm{N}(0, \sigma^2)$ and the weight vector $\boldsymbol{\lambda}$ satisfies the regularity conditions (9) and (11). By the theorem

$$E(\tilde{\boldsymbol{\beta}}) = \boldsymbol{\beta} - \mathbf{Q}\, E(\mathbf{W}\hat{\boldsymbol{\theta}} - \boldsymbol{\theta}), \qquad \mathrm{Var}(\tilde{\boldsymbol{\beta}}) = \sigma^2(\mathbf{X}'\mathbf{X})^{-1} + \mathbf{Q}[\mathrm{Var}(\mathbf{W}\hat{\boldsymbol{\theta}})]\mathbf{Q}'$$

and hence

$$MSE(\tilde{\boldsymbol{\beta}}) = \sigma^2(\mathbf{X}'\mathbf{X})^{-1} + \mathbf{Q}[MSE(\mathbf{W}\hat{\boldsymbol{\theta}})]\mathbf{Q}'.$$

The major ingredient of the proof is that the estimator $\hat{\boldsymbol{\theta}}$ in (10) and $\hat{\boldsymbol{\beta}}_0$ are uncorrelated and under the normality assumption they are independent. Now the relatively simple estimator $\mathbf{W}\hat{\boldsymbol{\theta}}$ of $\boldsymbol{\theta}$ characterizes the important features of the more complicated WALS estimator $\tilde{\boldsymbol{\beta}}$ of $\boldsymbol{\beta}$.

There is a growing literature on MA, see Hoeting et al.[13] for a review of Bayesian methods, and Claeskens and Hjort[14] on frequentist methods. Hansen[10] and Hansen and Racine,[15] for example, have developed methods to estimate the model weights in view of reducing estimation variance while controlling omitted variables bias. In practice the number of weights to be estimated can be huge, and therefore the set of candidate models is usually restricted to a small fraction of all possible models. However, the effect of this "preselection" is usually ignored.

We assume the approach proposed by Magnus et al.[11] where instead of every single weight λ_i we estimate the diagonal elements (12) of $\mathbf{W}$. Then the core of the WALS estimator (10) will be to find a good shrinkage estimator

$$\tilde{\boldsymbol{\theta}} = \mathbf{W}\hat{\boldsymbol{\theta}}, \quad 0 \leq |\tilde{\theta}_i| \leq |\hat{\theta}_i|,\ i = 1, \ldots, m, \tag{13}$$

of $\boldsymbol{\theta}$. Magnus et al.[11] assumed that each diagonal element $w_j = w_j(\hat{\theta}_j)$ depends only on $\hat{\theta}_j$, the jth element of $\hat{\boldsymbol{\theta}}, 1 \leq j \leq m$. Since $\hat{\theta}_1, \ldots, \hat{\theta}_m$ are independent under the normality assumption, also $w_1, \ldots, w_m$ are independent. Assuming that σ^2 is known, we have to find the best estimator of θ_j when $\hat{\theta}_j \sim \mathrm{N}(\theta_j, \sigma^2)$, $1 \leq j \leq m$. Thus we have m independent estimation problems. The case of unknown σ^2 will be discussed later. If the number of auxiliary regressors is large, say $m = 50$, then computing time of WALS is only of order 50. If estimation of every single weight $\lambda_i, 0 \leq i \leq M$ is required, the computing time will be of order 2^{50}. Thus the proposed WALS technique is computationally superior to techniques that require the estimation of every single weight.

4. Shrinkage with penalized LS

4.1. *Shrinkage estimation*

The essence of WALS estimation is the shrinkage estimator (13) of $\boldsymbol{\theta}$ presented in (10), where $\hat{\boldsymbol{\theta}}$ is the LS estimator of $\boldsymbol{\theta}$ and $\mathbf{W}$ is a random $m \times m$ diagonal matrix with diagonal elements $w_i, \quad 0 \le w_i \le 1,\ i = 1, \ldots, m$ (see (12)). Thus w_i's shrink the LS estimates $\hat{\theta}_i$ towards zero, and consequently $0 \le |\tilde{\theta}_i| \le |\hat{\theta}_i|,\ i = 1, \ldots, m$. Further, we assume that the shrinkage functions are even: $w_i(-\hat{\theta}_i) = w_i(\hat{\theta}_i),\ i = 1, \ldots, m$. Thus the functions $\tilde{\theta}_i$ are odd: $\tilde{\theta}_i(-\hat{\theta}_i) = -\tilde{\theta}_i(\hat{\theta}_i)$. Magnus et al.[11] and Einmahl et al.[16] adopted a Bayesian view on estimation deciding on to advocate the Laplace and Subbotin estimators which are of shrinkage type. The Laplace and Subbotin estimators are defined in Subsection 4.4.

The proposed estimators (13) are computationally superior to estimators that require estimation of every single weight in (9), since in estimation of $\tilde{\boldsymbol{\theta}}$ in (13) we have only m independent estimation problems $\tilde{\theta}_i = w_i \hat{\theta}_i$. We are now ready to define an important class of shrinkage estimators for θ. In the sequel $\mathcal{S}$ denotes this class and we call the estimators in $\mathcal{S}$ simply shrinkage estimators.

Definition 4.1. A real valued estimator δ of θ defined on $\mathbb{R}$ is a shrinkage estimator if the following four conditions hold:

(a) $0 \le \delta(\hat{\theta}) \le \hat{\theta} \quad$ for $\hat{\theta} \ge 0$,
(b) $\delta(-\hat{\theta}) = -\delta(\hat{\theta})$,
(c) $\delta(\hat{\theta})/\hat{\theta}$ is nondecreasing on $[0, \infty)$ and
(d) $\delta(\hat{\theta})$ is continuous,

where $\hat{\theta}$ is the LS estimator of θ.

In addition to shrinkage property (a) and antisymmetry (b), the definition puts two further requirements for shrinkage estimators. Consider now the condition (c). Denote $w(\hat{\theta}) = \delta(\hat{\theta})/\hat{\theta}$ for $\hat{\theta} > 0$ and think $\delta(\hat{\theta})$ as a weighted average of $\hat{\theta}$ and 0: $\delta(\hat{\theta}) = w(\hat{\theta})\hat{\theta} + (1 - w(\hat{\theta}))\,0$. The larger is $|\hat{\theta}|$, the better $\hat{\theta}$ is as an estimator of θ. Hence, when $\hat{\theta}$ increases we wish to put more weight on $\hat{\theta}$ than on 0, i.e., we wish to make $w(\hat{\theta})$ larger. Thus we see that the condition (c) makes sense. Condition (d) is a minimal smoothness condition which guarantees certain stability of estimation in the sense that small changes of data cannot create excessive variation of estimates.

4.2. *Penalized LS estimation*

Fitting the orthogonalized model (2) can be considered as a two-step least squares procedure.[1] The first step is to calculate $\hat{\beta}_0 = (\mathbf{X}'\mathbf{X})^{-1}\mathbf{X}'\boldsymbol{y}$ and replace $\boldsymbol{y}$ by $\boldsymbol{y} - \mathbf{X}\hat{\beta}_0 = \mathbf{M}\boldsymbol{y}$, where $\mathbf{M}$ is defined in (3). Then denote $\boldsymbol{z} = \mathbf{U}'\boldsymbol{y}$, and note that from the definition of $\mathbf{U}$ in (4) follows the equality$\mathbf{U}'\mathbf{M} = \mathbf{U}'$. Then the model $\mathcal{M}_M$ in (5) takes the form

$$\boldsymbol{z} = \boldsymbol{\theta} + \mathbf{U}'\varepsilon, \qquad \mathbf{U}'\varepsilon \sim (\mathbf{0}, \sigma^2\mathbf{I}_m). \tag{14}$$

The second step is to estimate $\boldsymbol{\theta}$ from the model (14).

In estimation of $\boldsymbol{\theta}$ we will use the penalized LS technique. If the penalty function satisfies proper regularity conditions, then the penalized LS yields a solution which is a shrinkage estimator of $\boldsymbol{\theta}$. In this approach we choose a suitable penalty function in order to get a shinkage estimator with good risk properties. The related Bayesian technique is to impose certain restrictions on the prior density, see e.g. Einmahl et al.[16] So, we are able to characterize a variety of interesting estimators from which many have already shown their potential in applications. This technique is also computationally efficient.

The penalized least squares estimate (PenLS) of $\boldsymbol{\theta} = (\theta_1, \ldots, \theta_m)'$ is the minimizer of

$$\frac{1}{2}\sum_{i=1}^{m}(z_i - \theta_i)^2 + \sum_{i=1}^{m} p_\lambda(|\theta_i|), \tag{15}$$

where $\lambda > 0$. It is assumed that the penalty function $p_\lambda(\cdot)$ is

$$\begin{aligned} &\text{(i) nonnegative,} \\ &\text{(ii) nondecreasing and} \\ &\text{(iii) differentiable on } [0, \infty). \end{aligned} \tag{16}$$

Minimization of (15) is equivalent to minimization componentwise. Thus we may simply minimize

$$l(\theta) = \frac{1}{2}(z - \theta)^2 + p_\lambda(|\theta|) \tag{17}$$

with respect to θ.

Example 4.1. There are close connections between the PenLS and variable selection or the PenLS and ridge regression, for example. Taking the L_2 penalty $p_\lambda(|\theta|) = \frac{\lambda}{2}|\theta|^2$ yields the ridge estimator

$$\check{\theta}_R = \frac{1}{1+\rho} z,$$

where $\rho > 0$ depends on λ. The hard thresholding penalty function

$$p_\lambda(|\theta|) = \lambda^2 - \frac{1}{2}(|\theta| - \lambda)^2(\mathbb{I}(|\theta| < \lambda)$$

yields the hard thresholding rule

$$\breve{\theta}_H = z\,\{\mathbb{I}(|z| > \lambda)\}, \tag{18}$$

where $\mathbb{I}(\cdot)$ is the indicator function. Then the minimizer of the expression (15) is $z_j\{\mathbb{I}(|\theta_j| > \lambda)\}$, $j = 1, \ldots, m$, and it coincides with the best subset selection for orthonormal designs. In statistics (see e.g. Morris et al.[17]) and in econometrics (see, e.g. Judge *et al.*[18]), the hard thresholding rule is traditionally called the pretest estimator.

The following theorem gives sufficient conditions for the PenLS estimate $\breve{\theta}$ of θ to be a shrinkage estimator. Further, the theorem provides the lower bound of the mean squared error

$$MSE(\theta, \breve{\theta}) = E[\breve{\theta}(z) - \theta]^2 = \mathrm{Var}[\breve{\theta}(z)] + \mathrm{Bias}(\theta, \breve{\theta}),$$

where $\mathrm{Bias}(\theta, \breve{\theta}) = \{E[\breve{\theta}(z)] - \theta\}^2$. This lower bound is called the *efficiency bound.*

Theorem 4.1. *We assume that the penalty function $p_\lambda(\cdot)$ satisfies the assumptions* (16). *We make two assertions.*

(i) If the three conditions hold

(1) the function $-\theta - p'_\lambda(\theta)$ is strictly unimodal on $[0, \infty)$,
(2) $p'_\lambda(\cdot)$ is continuous and nonincreasing on $[0, \infty)$, and
(3) $\min_\theta\{|\theta| + p'_\lambda(|\theta|)\} = p'_\lambda(0)$,

then the PenLS estimate $\breve{\theta}$ of θ belongs to the shrinkage family $\mathcal{S}$.

(ii) If the conditions of the assertion (i) hold and z follows the normal distribution $\mathrm{N}(0, \sigma^2)$, where σ^2 is known, the efficiency bound of $\breve{\theta}$ is

$$\inf_{\breve{\theta} \in \mathcal{S}} MSE(\theta, \breve{\theta}) = \frac{\theta^2}{1 + \theta^2}.$$

Proof. (i) The derivative $l'(\theta)$ of the function $l(\theta)$ to be minimized in (17) is

$$l'(\theta) = \mathrm{sgn}(\theta)\{|\theta| + p'_\lambda(|\theta|)\} - z.$$

If the three conditions in (i) hold, then by Theorem 1 in Antoniadis and Fan[19] the solution to the minimization problem (17) exists, is unique and takes the form

$$\breve{\theta}(z) = \begin{cases} 0, & \text{if } |z| \leq p_0, \\ z - \text{sgn}(z)\, p'_\lambda(|z|), & \text{if } |z| > p_0, \end{cases} \tag{19}$$

where $p_0 = \min_{\theta \geq 0}\{\theta + p'_\lambda(\theta)\}$. Clearly the solution (19) is antisymmetric, i.e. $\breve{\theta}(-z) = -\breve{\theta}(z)$. Since $p'_\lambda(z) \geq 0$ for $z \geq 0$, $\breve{\theta}(z)$ satisfies the shrinkage property (a) of definition 4.1: $0 \leq \breve{\theta}(z) \leq z$ for $z \geq 0$.

If $\min_\theta\{|\theta| + p'_\lambda(|\theta|)\} = p'_\lambda(0)$, then $p_0 = p'_\lambda(0)\}$ and the PenLS estimator (19) is continuous. Furthermore, since $p'_\lambda(\cdot)$ is nonincreasing on $[0, \infty)$, it follows that $\breve{\theta}(z)/z$ defined by (19) is nondecreasing on $[0, \infty)$. Hence the estimator (19) fulfils the condition (c) in Definition 4.1. Thus we have proved that the PenLS estimator (19) belongs to the shrinkage class $\mathcal{S}$.

(ii) By the assertion (i) the PenLS estimator $\breve{\theta}(z)$ belongs to shrinkage family $\mathcal{S}$, and consequently $\breve{\theta}(z)$ satisfies the regularity conditions R1 in Magnus:[20]

(a) $0 \leq \breve{\theta}(z)/z \leq 1$ for all z,
(b) $\breve{\theta}(-z)/(-z) = \breve{\theta}(z)/z$ for all z,
(c) $\breve{\theta}(z)/z$ is nondecreasing on $[0, \infty)$ and
(d) $\breve{\theta}(z)/z$ is continuous.

Hence by Theorem A7 in Magnus[20] the efficiency bound for the shrinkage estimators $\mathcal{S}$ is

$$\inf_{\breve{\theta} \in \mathcal{S}} MSE(\theta, \breve{\theta}) = \frac{\theta^2}{1 + \theta^2}.$$

This concludes the proof of the theorem. □

Note that the pretest estimator $\breve{\theta}_H$ given in (18) is not continuous, and hence it does not belong to the class of shrinkage estimators $\mathcal{S}$. Magnus[21] demonstrates a number of undesiderable properties of the pretest estimator. It is inadmissible and there is a range of values for which the MSE of $\breve{\theta}_H$ is greater than the MSE of both the least squares estimator $\hat{\theta}(z) = z$ and the null estimator $\hat{\theta}(z) \equiv 0$. The traditional pretest at the usual 5% level of significance results in an estimator that is close to having worst possible performance with respect to the MSE criterion in the neighborhood of the value $|\theta/\sigma| = 1$ which was shown to be of crucial importance.

Example 4.2. The L_q penalty $p_\lambda(|\theta|) = \lambda\,|\theta|^q$, $q \geq 0$ results in a bridge regression.[22] The derivative $p'_\lambda(\cdot)$ of the L_q penalty is nonincreasing on $[0, \infty)$ only when $q \leq 1$ and the solution is continuous only when $q \geq 1$.

Therefore, only L_1 penalty in this family yields a shrinkage estimator. This estimator is a soft thresholding rule, proposed by Donoho and Johnstone,[23]

$$\breve{\theta}_S = \text{sgn}(z)(|z| - \lambda)_+, \tag{20}$$

where z_+ is shorthand for $\max\{z, 0\}$. LASSO[24] is the PenLS estimate with the L_1 penalty in the general least squares and likelihood settings.

Since we have the efficiency bound of the PenLS estimators (19), the *regret* of $\breve{\theta}(z)$ can be defined as

$$r(\theta, \breve{\theta}) = MSE(\theta, \breve{\theta}) - \frac{\theta^2}{1+\theta^2}.$$

We wish to find an estimator with the desirable property that its risk is uniformly close to the infeasible efficiency bound. In search of such an estimator we may adopt the minimax regret criterion where we minimize the maximum regret instead of the maximum risk. An estimator $\breve{\theta}^*$ is *minimax regret* if

$$\sup_{\theta} r(\theta, \breve{\theta}^*) = \inf_{\breve{\theta} \in \mathcal{S}} \sup_{\theta} r(\theta, \breve{\theta}).$$

In theoretical considerations σ^2 is assumed to be known, and hence we can always consider the variable z/σ. Then expectation E is simply taken with respect to the $N(\theta, 1)$ distribution, and comparison of estimators risk performance is done under this assumption. In practical applications we replace the unknown σ^2 with s^2, the estimate in the unrestricted model. Danilov[12] demonstrated that effects of estimating σ^2 are small in case of Laplace estimator. We expect the approximation to be accurate for other shrinkage estimators too, although more work is needed to clarify this issue.

4.3. *Good PenLS shrinkage estimators*

In this subsection we consider properties of three well known PenLS estimators which are shrinkage estimators. The performance of two of them is also displayed in simulation experiments. Bruce and Gao[25] compared hard and soft thresholding rules and showed that hard thresholding tends to have bigger variance whereas soft thresholding tends to have bigger bias. To remedy the drawbacks of hard and soft thresholding, Fan and Li[26] suggested using continuous differentiable penalty function defined by

$$p'_\lambda(|\theta|) = \lambda \left\{ \mathbb{I}(|\theta| \le \lambda) + \frac{(a\lambda - |\theta|)_+}{(a-1)\lambda} \mathbb{I}(|\theta| > \lambda) \right\} \tag{21}$$

for some $a > 2$ and $\theta > 0$. The penalty (21) is called *smoothly clipped absolute deviation* (SCAD) penalty. Note that if the penalty function in (15) is constant, i.e. $p'(|\theta|) = 0$, then the rule in (19) takes the form $\hat{\theta}(z) \equiv z$ which is unbiased. Since the SCAD penalty $p'_\lambda(\theta) = 0$ for $\theta > a\lambda$, the resulting solution (Fan and Li[26])

$$\check{\theta}_{scad}(z) = \begin{cases} \text{sgn}\,(z)(|z| - \lambda)_+, & \text{if } |z| \le 2\,\lambda, \\ \frac{(a-1)z - \text{sgn}\,(z)a\lambda}{(a-2)}, & \text{if } 2\lambda < |z| \le a\lambda, \\ z, & \text{if } |z| > a\lambda \end{cases} \tag{22}$$

tends to be unbiased for large values of z. This estimator (22) can be viewed as a combination of soft thresholding for "small" $|z|$ and hard thresholding for "large" $|z|$, with a piecewise linear interpolation inbetween.

The SCAD estimator is closely related to the firm threshholding rule of Bruce and Gao:[25]

$$\check{\theta}_F(z) = \begin{cases} 0, & \text{if } |z| \le \lambda_1, \\ \text{sgn}(z)\,\frac{\lambda_2(|z| - \lambda_1)}{\lambda_2 - \lambda_1}, & \text{if } \lambda_1 < |z| \le \lambda_2, \\ z, & \text{if } |z| > \lambda_2, \end{cases} \tag{23}$$

where $0 < \lambda_1 < \lambda_2$. This rule was also suggested to ameliorate the drawbacks of hard and soft thresholding . For soft thresholding $p'(|\theta|) = \lambda$ for all θ, and $\hat{\theta}_S$ is biased also for large values of $|z|$. Bruce and Gao[25] showed that $MSE(\theta, \hat{\theta}_S) \to 1 + \lambda^2$ as $\theta \to \infty$ whereas $MSE(\theta, \hat{\theta}_F) \to 1$ as $\theta \to \infty$ (Bruce and Gao[25]) when $\lambda_2 < \infty$.

Breiman[27] applied the non-negative garrote rule

$$\check{\theta}_G(z) = \begin{cases} 0, & \text{if } |z| \le \lambda, \\ z - \lambda^2/z, & \text{if } |z| > \lambda \end{cases} \tag{24}$$

to subset selection in regression to overcome the drawbacks of stepwise variable selection rule and ridge regression. The MSE for the estimator $\hat{\theta}_G$ is comparable to that for the firm thresholding rule.[25,28] It is straightforward to show that the soft thresholding (20), SCAD (22), firm thresholding (23) and non-negative garrote (24) estimators belong to the shrinkage class $\mathcal{S}$ (Definition 4.1). The usual LS estimator $\hat{\theta}(z) \equiv z$ is a good candidate for large z, and hence we wish that for large z an estimator $\check{\theta}(z)$ is close to z in the sense that $z - \check{\theta}(z)$ converges to zero. It can be readily seen that the estimators $\check{\theta}_{scad}, \check{\theta}_F$ and $\check{\theta}_G$ have this property, i.e. $z - \check{\theta}(z) \to 0$ as $z \to \infty$ when $\check{\theta}(z)$ is any of the foregoing three estimators. For the soft thresholding rule $z - \check{\theta}_S(z)$ converges to a positive constant, but not to zero.

4.4. *The Laplace and Subbotin estimators*

Magnus[20] addressed the question of finding an estimator of θ which is admissible, has bounded risk, has good risk performance around $\theta = 1$, and is optimal or near optimal in terms of minimax regret when $z \sim \mathrm{N}(\theta, 1)$. The Laplace estimator

$$\hat{\theta}_L(z) = z - h(y)c$$

proved to be such an estimator, when $c = log\, 2$ and $h(\cdot)$ is a given antisymmetric monotonically increasing function on $(-\infty, \infty)$ with $h(0) = 0$ and $h(\infty) = 1$. The Laplace estimator is the mean of the posterior distribution of $\theta|z$ when a Laplace prior for θ with median$(\theta) = 0$ and median$(\theta^2) = 1$ is assumed. In search of prior which appropriately reflects the notion of ignorance, Einmahl et al.[16] arrived at the Subbotin prior that belongs to the class of reflected gamma densities. In practical applications they recommended the Subbotin prior

$$\pi(\theta) = \frac{c^2}{4} e^{-c|\theta|^{1/2}}$$

with $c = 1.6783$ which should stay close to the Laplace prior. Magnus et al.[11] and Einmahl et al.[16] also showed that the computational burden of the Laplace and Subbotin estimators is light when applied in the context of weighted average least squares (WALS). In our simulation experiments we compare the performance of these two Bayesian estimators, the Laplace and Subbotin, with the performance of the penalized LS estimators.

4.5. *Implementation using penalized LS*

We now recap the main steps of the penalized LS estimation of the parameters $\boldsymbol{\beta}$ and $\boldsymbol{\gamma}$ in the model (1). To orthogonalize the model (1) fix a matrix $\mathbf{C}$ such that $\mathbf{C}'\mathbf{Z}'\mathbf{MZC} = \mathbf{I}_m$. We can use the spectral decomposition $\mathbf{Z}'\mathbf{MZ} = \mathbf{P\Phi P}$' of $\mathbf{Z}'\mathbf{MZ}$ to have $\mathbf{C} = \mathbf{P\Phi}^{-1/2}$, where $\mathbf{\Phi} = \mathrm{diag}(\phi_1, \ldots, \phi_m)$ is the diagonal matrix of the eigenvalues of $\mathbf{Z}'\mathbf{MZ}$ and the columns of $\mathbf{P}$ is an orthonormal set of eigenvectors corresponding to these eigenvalues.

(1) Compute $\boldsymbol{y} - \mathbf{X}\hat{\boldsymbol{\beta}}_0 = \mathbf{M}\boldsymbol{y}$ and $\hat{\boldsymbol{\theta}} = \mathbf{C}'\mathbf{Z}'\mathbf{M}\boldsymbol{y}$, where $\mathbf{M} = \mathbf{I}_n - \mathbf{X}(\mathbf{X}'\mathbf{X})^{-1}\mathbf{X}'$.
(2) Compute $\hat{\underline{\boldsymbol{\theta}}} := \hat{\boldsymbol{\theta}}/\sigma$, where $\underline{\boldsymbol{\theta}}$ denotes $\boldsymbol{\theta}/\sigma$ and σ^2 is assumed to be known.
(3) For $j = 1, \ldots, m$ compute the PenLS estimate $\breve{\underline{\theta}}_j$ and its variance $\breve{\omega}_j^2$. Denote $\breve{\underline{\boldsymbol{\theta}}} = (\breve{\underline{\theta}}_1, \ldots, \breve{\underline{\theta}}_m)'$ and $\breve{\mathbf{\Omega}} = \mathrm{diag}(\breve{\omega}_1^2, \ldots, \breve{\omega}_m^2)$.

(4) The PenLS estimates for $\boldsymbol{\gamma}$ and $\boldsymbol{\beta}$ are

$$\breve{\gamma} = \sigma \mathbf{C}\underline{\breve{\boldsymbol{\theta}}} \quad \text{and} \quad \breve{\boldsymbol{\beta}} = (\mathbf{X}'\mathbf{X})^{-1}\mathbf{X}'(\boldsymbol{y} - \mathbf{Z}\breve{\gamma}),$$

since $\boldsymbol{\gamma} = \mathbf{C}\boldsymbol{\theta}$ and $\boldsymbol{\theta} = \sigma\underline{\boldsymbol{\theta}}$.

(5) The variance for $\breve{\gamma}$ and $\breve{\boldsymbol{\beta}}$ are

$$\begin{aligned} \mathrm{Var}(\breve{\gamma}) &= \sigma^2 \mathbf{C}\breve{\boldsymbol{\Omega}}\mathbf{C}' \\ \mathrm{Var}(\breve{\boldsymbol{\beta}}) &= \sigma^2(\mathbf{X}'\mathbf{X})^{-1} + \underline{\mathbf{Q}}\,\mathrm{var}(\breve{\gamma})\underline{\mathbf{Q}}', \end{aligned}$$

where $\underline{\mathbf{Q}} = (\mathbf{X}'\mathbf{X})^{-1}\mathbf{X}'\mathbf{Z}$. Finally we have $\mathrm{Cov}(\breve{\boldsymbol{\beta}}, \breve{\gamma}) = -\underline{\mathbf{Q}}\,\mathrm{var}(\breve{\gamma})$,.

In practice σ^2 is unknown and it is replaced with s^2, the sample variance estimated in the unrestricted model.

5. The costs of initial hospitalization for a first hip fracture

We compare the estimation techniques presented in this paper on hip fracture data. The original purpose of our dataset is to compare treatment costs of hip fracture patients between hospital districts in Finland. In this paper we use it to demonstrate the performance of various penalized least squares estimators.

The dataset was obtained by combining data from several national registries.[29] The costs of the first institutionalization period of first time hip fracture patients in Finland were calculated in the time period of 1999 – 2005. There are a total of 21 hospital districts in Finland, but in the estimations in this paper we are only using the seven largest districts. The dataset was made more homogenous by keeping such patients in the data who had not been institutionalized before the fracture and who were not institutionalized after the fracture either. Patients who died within a year after the fracture were removed. The final dataset used in this paper contained 11961 patients of age 50 or older.

As the dependent variable in our model we are using the cost of the first continuous instituionalization period. In our model we have 7 focus regressors, which are dummy variables for the six largest hospital districts and 31 auxilary regressors. The largest hospital district was taken as the baseline. The set of auxilary regressors contains information on the patients such as gender, age and time between fracture and operation and a number of important comorbidities like congestive heart failure, diabetes and cancer. The auxilary regressors are intended to reflect the mix of patients treated in a hospital district.

6. Simulation experiments

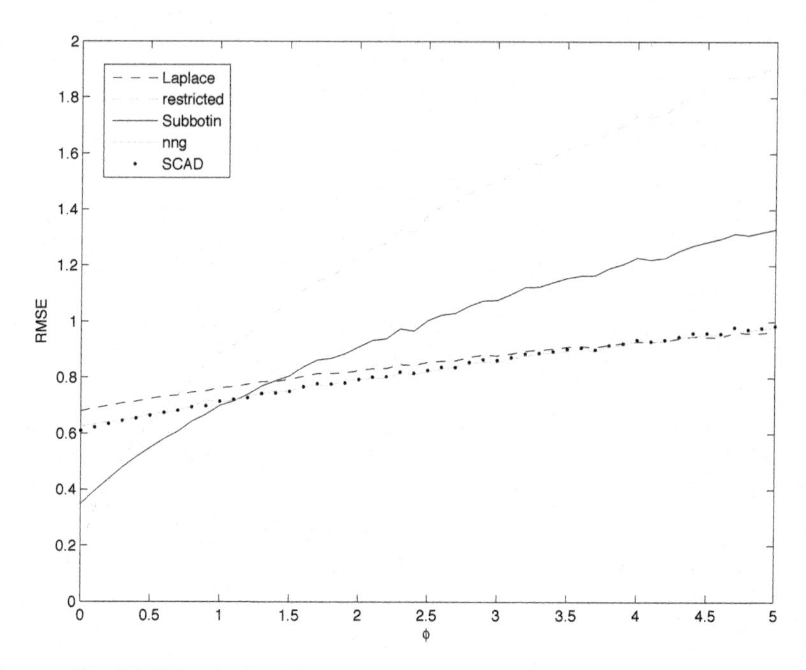

Fig. 1. The RMSE values of the Laplace, non-negative garrote (nng), restricted LS, SCAD and Subbotin estimators are compared to the unrestricted model. $RMSE = 1$ is the $RMSE$-value of the unrestricted model.

The purpose of the simulations is to compare the performance of various PenLS estimators, including the restricted LS estimator, with the performance of the Laplace estimator within a realistic set-up. The Laplace estimator has been shown to be theoretically and practically superior to many existing MA methods.[11,20] Recently Einmahl et al.[16] proposed a competitor for it, the Subbotin estimator. Therefore also the Subbotin estimator is included in our simulation study.

We use the LS estimator in the unrestricted model as our benchmark. We take the estimates from the unrestricted model as the 'true' parameter values. We do not generate the disturbances from a theoretical distribution, but the disturbances are obtained by resampling the LS residuals of the estimated unrestricted model. Thus the simulations are based on real data, not on generated data. The disturbances in each round of the simulation experiment are obtained by randomly selecting 2000 numbers with replacement from the LS residuals. In order to gain broader perception of the estimators performance we use different values of γ by scaling it. This is carried out so that we replace γ by $\tau\gamma$ where the scale factor τ is obtained from the equality

$$\phi = \tau^2\gamma'\mathbf{Z}'\mathbf{M}\mathbf{Z}\gamma,$$

when we let ϕ vary between 0 and 5. Here ϕ can be considered the approximation of the theoretical F-ratio $\gamma'\mathbf{Z}'\mathbf{M}\mathbf{Z}\gamma/(m\sigma^2)$.

We concentrate on the performance of our focus parameters, the β-parameters. Based on 10000 replications we approximate the distribution of $\hat{\boldsymbol{\beta}}$. The estimators are evaluated in terms of the root mean squared error ($RMSE$). Let $\hat{\boldsymbol{\beta}}^{(i)}$ denote the estimate of $\boldsymbol{\beta}$ in the i-th replication, and we compute

$$RMSE(\hat{\boldsymbol{\beta}}) = \sqrt{\sum_{i=1}^{N} \|\hat{\boldsymbol{\beta}}^{(i)} - \boldsymbol{\beta}\|^2/N},$$

where $\|\cdot\|$ denotes the Euclidean norm, $N = 10000$ is the number of replicates and $\boldsymbol{\beta}$ is the estimate from the unrestricted model. The $RMSE$ of each estimator is computed. Since the LS estimator in the unrestricted model is used as the benchmark, the $RMSE$ of an estimator is divided by the $RMSE$ computed from the unrestricted model. So, $RMSE = 1$ in In Figure 1 means that the $RMSE$ of an estimator is equal to that of the unrestricted model.

The parameter values of the SCAD and the non-negative garrote were chosen so that the theoretical risk (MSE) of the estimators are uniformly close to the efficiency bound of the shrinkage estimators. For SCAD we used parameter values $a = 5$ and $\lambda = 0.5$ and for the non-negative garrote we take $\lambda = 0.01$. For these parameter values the MSE of the SCAD and the non-negative garrote were also close to the MSE of the Laplace estimator.

In Figure 1 we have compared the $RMSE$'s of the competing estimators as ϕ increases. We observe that the Laplace estimator and SCAD perform better than the unrestricted model with all ϕ values. The SCAD estimator does a little better than Laplace with small and intermediate ϕ values. The non-negative garrote estimator performs equally well with SCAD. Subbotin performs very well with $\phi < 1$, but with larger ϕ values loses to SCAD, Laplace, non-negative garrote and the unrestricted model.

7. Concluding remarks

In model selection one attempts to use the data to find a single "winning" model, according to a given criterion, whereas with model averaging (MA) one seeks a smooth compromise across a set of competing models. Most existing MA methods are based on estimation of all model weights using exponential Akaike information criterion (AIC) or Bayesian information criterion (BIC) weights, for example. A common challenge for a regression analyst is the selection of the best subset from a set of m predictor variables in terms of some specified criterion. Then the number of competing models is 2^m, and consequently the computational burden to estimate all the model weights becomes soon too heavy when m is large.

The quality of the WALS (10) estimator depends on the shrinkage estimator of the auxiliary parameter γ where each shrinkage factor is a sum of model weights. So, estimation of 2^m model weights is converted into estimation of m shrinkage factors with trivial computational burden. We define the class of shrinkage estimators in view of MA and show that these shrinkage estimators can be constructed by putting appropriate restrictions on the penalty function. Utilizing the relationship between shrinkage and parameter penalization, we are able to build up computationally efficient MA estimators which are easy to implement. These estimators include some known recent contributions, like the non-negative garrote of Breiman,[27] the lasso-type estimator of Tibshirani[24] and the SCAD estimator of Fan and Li.[26] In the simulation experiments we assess the quality of an estimator in terms of its $RMSE$. In this competition the winners were the SCAD and non-negative garrote but the Laplace estimator did almost as well.

References

1. G. A. F. Seber, *Linear Regression Analysis* (Wiley, New York, 1977).
2. R. A. Horn and C. R. Johnson, *Matrix Analysis* (Cambridge University Press, Cambridge, 1985).
3. G. A. F. Seber and A. J. Lee, *Linear Regression Analysis*, 2nd edn. (Wiley, New York, 2003).
4. H. Akaike, Information theory as an extension of the maximum likelihood principle. Pages 267–281 *in* B. N. Petrov, and F. Csaki, eds. (Second International Symposium on Information Theory, Akademiai Kiado, Budapest, 1973).
5. G. Schwarz, Estimating the dimension of a model. *Annals of Statistics*, 6, 461–464 (1978).
6. J. Rissanen, Modeling by Shortest Data Description. *Automatica*, 14, No. 1, 465–471 (1978).
7. J. Rissanen, *Information and Complexity in Statistical Modeling.* (Springer, New York, 2007).
8. J. Rissanen, *Optimal Parameter Estimation.* (Cambridge University Press, Cambridge, 2011).
9. A. Miller, *Subset Selection in Regression.* (Chapman & Hall, Boca Raton, 2002).

10. B. E. Hansen Least squares model averaging. *Econometrika*, 75, 1175–1189 (2007).
11. J. R. Magnus, O. Powell, and P. Prüfer, A comparison of two model averaging techniques with an application to growth empirics. *Journal of Econometrics*, 154, 139–153 (2010).
12. D. Danilov, and J. R. Magnus, On the harm that ignoring pretesting can cause. *Journal of Econometrics,* 122, 27–46 (2004).
13. J. A. Hoeting, D. Madigan, A. E. Raftery and C. T. Volinsky, Bayesian model averaging: A tutorial (with discussion). *Statistical Science*, 14, 382–417 (1999).
14. N. I. Hjort, and G. Claeskens, Frequentist model averaging estimators. *Journal of the American statistical Association*, 98, 879–899 (2003).
15. B. E. Hansen and J. Racine, Jackknife model averaging. *Journal of Econometrics*, forthcoming (2011).
16. J. H. J. Einmahl, K. Kumar and J. R. Magnus Bayesian model averaging and the choice of prior. *CentER Discussion Paper*, No. 2011–003 (2011).
17. C. Morris, R. Radhakrishnan and S. L. Sclove, Nonoptimality of preliminary test estimators for the mean of a multivariate normals distribution. *Annals of Mathematical Statistics*, 43, 1481–1490 (1972).
18. G. G. Judge, W. E. Griffiths, R. C. Hill, H. Lutkepohl and T. C. Lee, *The Theory and Practice of Econometrics*, (Wiley, New York, 1985).
19. A. Antoniadis and J. Fan, Regularization of Wavelets Approximations. *Journal of the American statistical Association*, 96, 939–967 (2001).
20. J. R. Magnus, Estimation of the mean of a univariate normal distribution with a known variance. *Econometrics Journal,* 5, 225–236 (2002).
21. J. R. Magnus, The traditional pretest estimator. Theory of Probability and Its Applications, 44, 293–308 (1999).
22. I. E. Frank and J. H. Friedman, A statistical view of some chemometrics regression tools. *Technometrics*, 35, 109–148 (1993).
23. D. L. Donoho and I. M. Johnstone, Ideal spatial adaptation by wavelet shrinkage. *Biometrika*, 81, 425–456 (1994).
24. R. Tibshirani, Regression shrinkage and selection via the Lasso. *Journal of the Royal Statistical Society B*, 1, 267–288 (1996).
25. A. G. Bruce and H.-Y. Gao, Understanding WaveShrink: Variance and bias estimation. *Biometrica*, 83, 727–745 (1996).
26. J. Fan and R. Li, Variable Selection via Nonconcave Penalized Likelihood and Its Oracle Properties. *Journal of the American statistical Association*, 96, 1348–1360 (2001).
27. L. Breiman, Better subset regression using nonnegative garrote. *Technometrics*, 37, 373–384 (1995).
28. H.-Y. Gao, Wavelet Shrinkage Denoising Using the Non-Negative Garrote *Journal of Computational and Graphical Statistics,* 7, 469–488 (1998).
29. R. Sund, M. Juntunen, P. Lüthje, T. Huusko, M. Mäkelä, M. Linna, A. Liski, U. Häkkinen, *PERFECT - Hip Fracture, Performance, Effectiveness and Cost of Hip Fracture Treatment Episodes* (In Finnish), National Research and Development Centre for Welfare and Health, Helsinki, 2008.

K-NEAREST NEIGHBORS AS PRICING TOOL IN INSURANCE: A COMPARATIVE STUDY

K. PÄRNA, R. KANGRO, A. KAASIK and M. MÖLS

Institute of Mathematical Statistics, University of Tartu,
Tartu 50409, Estonia
E-mail: kalev.parna@ut.ee

The method of k-nearest neighbors (k-NN) is used for estimation of conditional expectation (regression) of an output Y given the value of an input vector $\boldsymbol{x}$. Such a regression problem arises, for example, in insurance where the pure premium for a new client (policy) $\boldsymbol{x}$ is to be found as conditional mean of the loss. In accordance with supervised learning set-up, a training set is assumed. We apply the k-NN method to a real data set by proposing solutions for feature weighting, distance weighting, and the choice of k. All the optimization procedures are based on cross-validation techniques. Comparisons with other methods of estimation of the regression function like regression trees and generalized linear models (quasi-Poisson regression) are drawn, demonstrating high competitiveness of the k-NN method.

Keywords: distance measures, feature selection, quasi-Poisson regression, k-nearest neighbors, local regression, premium calculation, regression tree

1. Introduction

The method of k-nearest neighbors (k-NN) is recognized as a simple but powerful toolkit in statistical learning.[1,2] It can be used for both discrete or continuous decision making, known as classification and regression, respectively. In the latter case the k-NN is aimed at estimation of conditional expectation $y(\boldsymbol{x}) := E(Y|X = \boldsymbol{x})$ of an output Y given the value of an input vector $\boldsymbol{x} = (x_1, \ldots, x_m)$. In accordance with supervised learning set-up, a training set is given consisting of n pairs $(\boldsymbol{x}_i, y_i)$ and the problem is to estimate $y(\boldsymbol{x})$ for a new input $\boldsymbol{x}$. Among numerous other application areas, this is exactly the situation in insurance where the pure premium $y(\boldsymbol{x})$ for a new client (policy) $\boldsymbol{x}$ is to be determined. More specifically, the price of a policy consists of several components:

$$gross\ premium = pure\ premium + risk\ loading + other\ costs.$$

By definition, the pure premium is equal to the expected value of possible claims, $y(\boldsymbol{x}) = E(Y|X = \boldsymbol{x})$, with $\boldsymbol{x}$ representing the features of both, the client and the object insured (e.g. car).

Some typical features, the components of $\boldsymbol{x} = (x_1, \ldots, x_m)$, can be named here:
– the object: value, age, make, model, initial price, the duration of the contract, etc.
– the client: sex, age, old or new client, loss history (if exists), etc.

The basic problem in estimating conditional expectation is that, as a rule, the training set does not contain any other record with the same $\boldsymbol{x}$, and thus also the other data points must be used when estimating $y(\boldsymbol{x})$. Three typical approaches to cope with the problem are:
– linear model fit by least squares : uses *all* points in the training set, but makes huge assumptions about the distribution of the data;[3]
– classification and regression trees (CART): building an hierarchical binary classification of training data and using final classes as basis for prediction;[4]
– k-nearest neighbors: uses only 'closest' points and makes only very mild assumptions.

Using the k-NN method, one first finds a neighborhood $U_{\boldsymbol{x}}$ consisting of k samples which are nearest to $\boldsymbol{x}$ with respect to a given distance measure d. Secondly, the average (or weighted average) of Y is calculated over the neighborhood $U_{\boldsymbol{x}}$ as an estimate of $y(\boldsymbol{x})$.[1] Although a natural idea and successfully applied in many areas, the k-NN does not seem to be widely used in insurance practice, at least not in a well documented form. One related paper is Ref. 5 where an attempt has been made to use k-NN in building an expert system for insurance premium assessment, which has some common features with our approach here.

In this paper all three aforementioned approaches are used as candidate methodologies for premium calculation in insurance. The methods have been tested on a real data set and (perhaps somewhat surprisingly) the k-NN has shown the best performance. Accordingly, we describe the k-NN method in a more detailed manner as compared to other methods.

The paper is organized as follows. We first describe the k-NN method and then analyze the key issues related to its application: the feature selection and weighting, distance weighting, and finding the optimum value of the smoothing parameter k. Secondly, two other candidate methods for estimation of the regression function $y(\boldsymbol{x})$, namely CART and quasi-Poisson regression (a generalized linear model), are described in short. Finally, numerical comparisons between candidate methods are drawn. The methods were calibrated on the basis of the older part of the data (training data) by using 10-fold cross-validation techniques, and then the calibrated models were tested on the rest of the data with respect to their ability to predict future losses.

2. k-NN in practice: key issues

In accordance with the supervised learning set-up a training set consisting of n pairs $(\boldsymbol{x}_i, y_i)$ is given and the problem is to estimate $y(\boldsymbol{x})$ for a new input $\boldsymbol{x}$. Since there is typically at most one observation at any point $\boldsymbol{x}$ in the training set, the k–NN method proposes to use also the points that are close (in some sense) to the target point. The idea behind is that if two inputs are close to each other then respective losses have tendency to be similar as well. The k-NN model is often used in so-called 'collaborative recommendation' systems that provide users with personalized suggestions for products or services (e.g. Amazon).[6]

2.1. *Basic steps of k-NN*

The k-NN method consists of two basic steps:

(1) Find the neighborhood $U_{\boldsymbol{x}}$ consisting of k samples in the data that are nearest to $\boldsymbol{x}$ with respect to a given distance measure d.

(2) Calculate the (weighted) average of Y over the neighborhood $U_{\boldsymbol{x}}$ as an estimate of $y(\boldsymbol{x})$:

$$\hat{y}(\boldsymbol{x}) := \frac{1}{\sum_{i \in U_{\boldsymbol{x}}} \alpha_i} \sum_{i \in U_{\boldsymbol{x}}} \alpha_i \cdot y_i, \tag{1}$$

where the weights α_i are chosen so that the nearer neighbors contribute more to the average than the more distant ones.

In order to find the neighborhood $U_{\boldsymbol{x}}$, a distance measure must be specified. In our application, the distance measure between the instances $\boldsymbol{x}_i$ and $\boldsymbol{x}_{i'}$ in the form

$$d(\boldsymbol{x}_i, \boldsymbol{x}_{i'}) = \sum_{j=1}^{m} w_j \cdot d_j(x_{ij}, x_{i'j}), \tag{2}$$

is used, where w_j is the weight of the feature j and $d_j(x_{ij}, x_{i'j})$ is the discrepancy of $\boldsymbol{x}_i$ and $\boldsymbol{x}_{i'}$ along the feature j. For numerical features we propose to take

$$d_j(x_{ij}, x_{i'j}) = (x_{ij} - x_{i'j})^2 \tag{3}$$

and for categorical features

$$d_j(x_{ij}, x_{i'j}) = \mathbf{1}_{x_{ij} \neq x_{i'j}} = \begin{cases} 1, & \text{if } x_{ij} \neq x_{i'j}, \\ 0, & \text{otherwise.} \end{cases} \tag{4}$$

From the formulas above several issues are seen that must be solved when applying the k-NN method:

- feature weighting w_j (including feature selection),
- distance weighting α_i,
- finding optimum value for k,
- handling missing values.

We next consider the four topics one-by-one.

2.2. *Feature weighting*

As to the feature weighting, we propose a three-factor weighting scheme

$$w_j := w_j(1) \cdot w_j(2) \cdot w_j(3).$$

The first component eliminates the scale effect (normalization) and is defined as

$$w_j(1) = (\bar{d}_j)^{-1}, \tag{5}$$

where $\bar{d}_j$ is the mean discrepancy between the samples along the feature j,

$$\bar{d}_j = \frac{1}{n^2} \sum_{i=1}^{n} \sum_{i'=1}^{n} d_j(x_{ij}, x_{i'j}). \tag{6}$$

The second component accounts for the dependence between the feature j and Y – the stronger the relationship, the bigger the weight of j. A standard measure of dependence is the coefficient of determination R_j^2, thus we define

$$w_j(2) := R_j^2 = \frac{SSB_j}{SST}, \tag{7}$$

where SST measures the total variance of Y and SSB_j is the part of the variance that is accounted for by the feature j. We apply this formula for both categorical and numerical features. More precisely, let the values of the feature j be divided into classes $C_1, \ldots, C_{m_j}$. In the case of nominal j, each value of j defines a separate class. If j is numerical, we divide the range of j into $m_j \approx \sqrt{n}$ classes of equal length. Let $\bar{y}$ and $\bar{y}_k$ be the total average of Y and the class average of Y, respectively. The two sums of squares in (7) can now be obtained as

$$SST = \sum_{i=1}^{n} (y_i - \bar{y})^2, \quad SSB_j = \sum_{k=1}^{m_j} n_k (\bar{y}_k - \bar{y})^2,$$

where n_k is the number of cases in the class C_k.

Finally, the third component is the feature selection – a backward procedure is proposed for obtaining the best subset of features J^* and

$$w_j(3) := \mathbf{1}_{j \in J^*}. \tag{8}$$

For its central importance, we explain the feature selection in details. Let $\hat{y}(\boldsymbol{x}_i)$ be an estimator of $E(Y|X = \boldsymbol{x}_i)$ which uses all features in the feature set J (initially J consists of m features). Let $\hat{y}_{-j}(\boldsymbol{x}_i)$ be the same estimator without using the feature j. Corresponding model errors are denoted as

$$S = \sum_{i=1}^{n} \big(y_i - \hat{y}(\boldsymbol{x}_i)\big)^2, \quad S_{-j} = \sum_{i=1}^{n} \big(y_i - \hat{y}_{-j}(\boldsymbol{x}_i)\big)^2. \tag{9}$$

The basic idea is that if $S_{-j} < S$, then the exclusion of j from J increases the quality of the model.

An iterative procedure for finding an optimal feature set is the following.

Step 0: Take $J^0 = \{1, \ldots, m\}$.
Step k: For the set $J^k \subset \{1, \ldots, m\}$ do:

(1) find the predictions $\hat{y}(\boldsymbol{x}_i)$, $\hat{y}_{-j}(\boldsymbol{x}_i)$, $i = 1, \ldots, n$, $j \in J^k$;
(2) find the squared errors S and S_{-j} of predictions for each $j \in J^k$;
(3) find

$$j^k := \arg\max_{j \in J^k} (S - S_{-j}).$$

If $S - S_{-j^k} \geq 0$, then $J^{k+1} := J^k \setminus \{j^k\}$.
Finish: if $S - S_{-j^k} < 0$. The optimal set $J^* := J^k$.

After the optimal feature set J^* is found, the third weight component $w_j(3)$ can be calculated as in (8).

Note that the iterative procedure for feature selection is the most time consuming stage of the k-NN method. As a by-product, the curse of dimensionality problem is solved in a natural way.

2.3. *Distance weighting*

Having found the neighborhood $U_{\boldsymbol{x}}$ for the input $\boldsymbol{x}$, a natural idea is to assign weights to points in $U_{\boldsymbol{x}}$ that die off with the distance from the target point, i.e. the larger the distance $d(\boldsymbol{x}, \boldsymbol{x}_i)$ the smaller the weight α_i. A popular kernel for weighting is the tricubic kernel

$$\alpha_i = \left[1 - (\frac{d(\boldsymbol{x}, \boldsymbol{x}_i)}{R_{\boldsymbol{x}}})^3\right]^3, \tag{10}$$

where $R_{\boldsymbol{x}}$ is the distance to the most distant neighbor:

$$R_{\boldsymbol{x}} = \max\{d(\boldsymbol{x}, \boldsymbol{x}_i) : i \in U_{\boldsymbol{x}}\}.$$

The effectiveness of the distance weighting depends on the bias/variance tradeoff: for a fixed number of neighbors the distance weighting reduces the bias, but increases the variance of the prediction. In our computations, we have used both the simple and the weighted average.

2.4. *Optimizing k*

The number of neighbors, k is a smoothing parameter. Whereas a large k provides stable estimates but possibly a poor fit with data, a small k gives unstable estimates, although better fit. The basic idea here is to try different values of k, when calibrating models on the training set. We have used a forward strategy starting with $k = 50$ and doubling the value of k at each next step until there is no improvement in the overall measure of goodness of the model. Note that in our application, for each k the best feature subset J^* was selected using the stepwise procedure above.

2.5. *Handling missing data*

Missing data is yet another practical problem one usually faces when handling real data. If some of the x_{ij}'s are absent then the discrepancies $d_j(x_{ij}, x_{i'j})$ can not be computed. One possible remedy is the imputation procedure – replacing the missing value x_{ij} with the mean $\bar{x}_j$ (for example), or another value by using more sophisticated methods. However, doing this one can easily create neighborhoods that are too heterogeneous. Therefore we recommend a different approach: if $d_j(x_{ij}, x_{i'j})$ can not be computed (for x_{ij} or $x_{i'j}$ being unknown), replace $d_j(x_{ij}, x_{i'j})$ with the average distance $\bar{d}_j$ w.r.t the feature j, defined in (6).

3. Comparison with other methods

In order to compare the proposed method with some popular alternatives, we used real vehicle casco insurance data of an insurance provider.

In addition to the k-nearest neighbors method with and without distance weighting, we considered the following two methods for calculating the pure premium.

3.1. *Classification and Regression Trees*

The idea of Classification & Regression Trees is to provide a simple yet useful rule for prediction. This rule can be stated using the "if...else..." construction recursively. First the input vector $\boldsymbol{x} = (x_1, \ldots, x_m)$ is scanned to find the best predictor i.e. either a continuous variable or a factor that allows splitting the dataset into two classes so that the deviance is minimized. Then the two new obtained classes are split (possibly using a different predictor) and so on.[4]

In particular, the deviance of the model (tree) $D(T)$ is typically defined as the sum of squares

$$D(T) = \sum_{i=1}^{n} (y_i - \mu_{[i]})^2, \tag{11}$$

where $\mu_{[i]}$ average of that class where the i-th observation belongs to (according to the tree T). At each step we have thus to make two choices: to fix the class and to find a risk factor with the splitting rule. At the j-th step we have j possibilities for the choice of the class (the number of classes $|T| = j$)and m risk factors. The number of possible splitting rules depends on the number of values of the risk factor. The stopping rule of the recursive method is implemented by searching for such a set of classes which will decrease the deviance more than some fixed number. If there does not exist such a partition, we stay with the previous obtained classification. This idea can be described also through the notion of "price" of the tree

$$D_\alpha(T) = D(T) + \alpha|T|, \tag{12}$$

where adding one leaf to the tree costs α and the penalty parameter $\alpha > 0$. We want to minimize the deviance over all possible subtrees of the maximal tree T_∞ (that we would get when $\alpha = 0$) or in other words we want to find for fixed α such a tree T, a subtree of T_∞, for which $D_\alpha(T)$ is minimal. Optimal value of α is determined using cross validation.

In the end we thus have a model that maps every possible $\boldsymbol{x}$ to a class. When using the tree for prediction we set $\hat{y}(\boldsymbol{x})$ equal to the mean of the class $\boldsymbol{x}$ belongs to.

3.2. *Quasi-Poisson model*

The Quasi-Poisson model used for comparison is a generalized linear model - basically the Poisson regression model with an overdispersion parameter (c_d) included. It uses a log-link and assumes the variance to be proportional to the mean:

$$\ln(\mathrm{E}Y) = c_0 + c_1 X_1 + \ldots c_m X_m,$$
$$\mathrm{Var}(Y) = c_d \mathrm{E}Y.$$

Negative-binomial ($\mathrm{Var}Y = \mathrm{E}Y + \mathrm{E}Y^2/c_d$) models were also considered as an alternative, but seemed to fit badly for this particular dataset (and often experienced convergence problems).

Model selection was based on the quasi-Akaike Information Criterion (qAIC),[7] one prefers a quasi-Poisson model with the smallest qAIC value:

$$\mathrm{qAIC} = -2\frac{\log L_{\mathrm{Poisson}}}{c_d^*} + 2p,$$

where p is number of parameters, c_d^* is the overdispersion parameter estimated from the saturated model ($c_d^* = 106153$) and L_{Poisson} is the likelihood calculated as for an ordinary Poisson regression model.

First a stepwise model selection procedure was carried out to select a relevant subset of the main effects. If a continuous variable was included into the model, then also second-degree b-splines with knots based on quantiles of the independent variable (or quantiles of unique values, if a few values dominate the distribution) were considered to approximate the unknown functional relationship. The number of knots was chosen by minimizing the qAIC-value. After including the relevant main effects all second-order interactions between the included variables were also considered.

3.3. *Numerical results*

The dataset spanned four years and contained data for approximately 25000 policies. We divided it into two parts: first three years of "Historical Data" and the last year of "Future Data". Based on "Historical Data" we computed a performance measure for each method of calculating expected traffic losses by two 10-fold cross validations as follows:

(1) Two different random orderings of the policies of the "Historical Data" were generated. The same orderings were used for all methods.
(2) For each random ordering:

    (a) The dataset of policies was divided into 10 approximately equal parts.
    (b) For each part, the model was fitted by using the data for 9 remaining parts.
    (c) Using that model, the expected losses $\hat{y}_i$ for the policies of the current part (testset) were predicted and the squared error was computed.

(3) The root mean squared error (RMSE) was computed as overall measure of quality of the results:

$$\sqrt{\sum_{i \in \text{testset}} (y_i - \hat{y}_i)^2 / |\text{testset}|}.$$

As an additional measure of performance of our estimation methods, the following final test-procedure was carried through. First, the methods were calibrated on the basis of the whole "Historical Data" dataset; then the calibrated methods were used to predict the traffic losses for the policies of "Future Data", and finally the root mean squared error (RMSE) was computed for each method showing how well the models fitted on historical data predict future losses.

All methods started model selection from the same list of 12 candidate factors determine from the information in policies including driver's age (if available), car's age, car's estimated market value, car's make and model. The results are presented in Figure 1, where smaller values correspond to better results. As we can see, the cross validation results based on the historical data predict well the behavior of the methods in predicting future losses. Also, both versions of the k-nearest neighbors method (with and without distance weighting) outperform the alternative methods. Although the differences of the performance measures seem to be small, it should be taken into account that in the case of the insurance data the unpredictable part of the variance is naturally high in comparison to the square of the expected value (due to the heavy tail of the loss distributions), so a small reduction in the prediction error may imply significant improvement in the predictions of the conditional expectations.

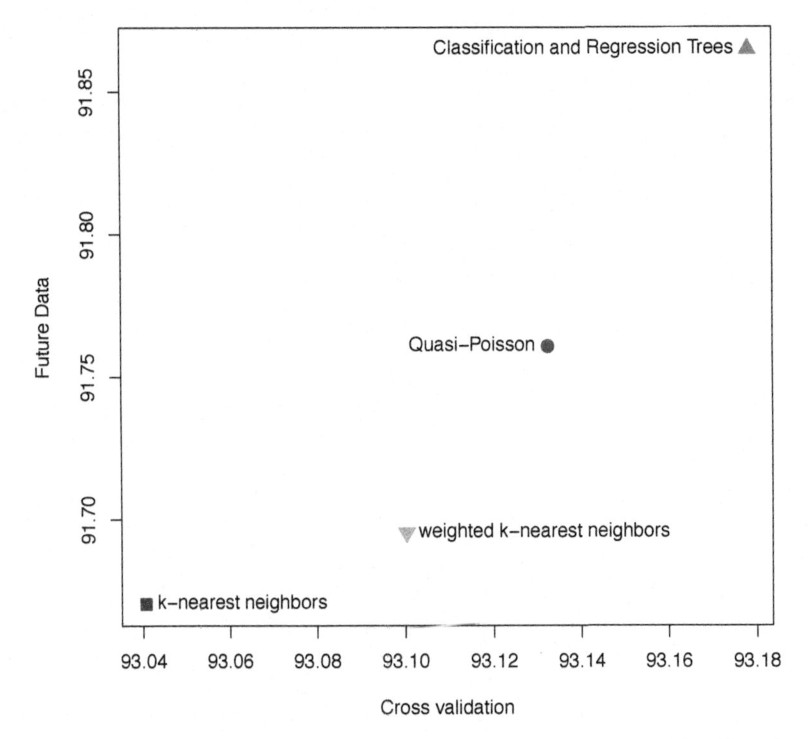

Fig. 1. Comparison of results of CART, Quasi-Poisson regression and k-NN with and without distance weighting.

4. Conclusions

Since the computational power of computers is continuously increasing, it has become possible to use methods for computing pure premiums that earlier were considered infeasible. We presented in this paper a method, that can be made fully automatic and dynamic (new data can immediately be taken into account in future predictions). Our final remarks about the k-nearest neighbors method are as follows:

- k-NN is highly competitive (at least in our application) and therefore is definitely worth considering as a possible premium calculation method in insurance.
- It is well known that for k-NN there is a bias problem at or near the boundaries (where the neighborhood only includes points from one side) [8, p. 73]. The possibilities of coping with this problem is a subject to further research.

- The method is computationally quite intensive but this is not a critical issue, especially since the method can easily be parallelized.
- It is algorithmic method and does not produce any explicit formula. This means that it is not easy to interpret the effects of different factors by analyzing the fitted model (but it is still possible to see the effects by producing appropriate graphs of predictions). On the other hand, the method does not require making assumptions about the type of dependence on different factors and therefore can approximate well quite complicated dependencies.

Acknowledgments

This research is supported by Targeted Financing Project SF0180015s12.

References

1. T. Hastie, R. Tibshirani and J. Friedman. *The Elements of Statistical Learning: Data Mining, Inference, and Prediction* (Springer, New York, 2001).
2. T. M. Mitchell. *Machine-Learning* (McGraw-Hill, Burr Ridge, 2001).
3. E. Ohlsson, B. Johansson. *Non-Life Insurance Pricing with Generalized Linear Models* (Springer, New York, 2010).
4. L. Breiman, J. Friedman, R. Olshen and C. Stone. *Classification and Regression Trees* (Chapman & Hall, New York, 1984).
5. S. Berkovsky, Y. Eytani and E. Furman. Developing a framework for insurance underwriting expert system. *IJSIT Lecture Notes of 1st International Conference on Informatics*, Vol.1, No.2, Sept., p.191–207 (2004).
6. Y. Koren. Factor in the neighbors: Scalable and accurate collaborative filtering. *ACM Transactions on Knowledge Discovery from Data (TKDD)*, Volume 4 Issue 1, January 2010., Article 1, 25p (2010).
7. D. R. Anderson, K. P. Burnham and G. C White. AIC model selection in overdispersed capture-recapture data. *Ecology*, **75**, 1780–1793 (1994).
8. L. Wasserman. *All of Nonparametric Statistics* (Springer, New York, 2006).

STATISTICAL STUDY OF FACTORS AFFECTING KNEE JOINT SPACE AND OSTEOPHYTES IN THE POPULATION WITH EARLY KNEE OSTEOARTHRITIS

T. von ROSEN[1], A. E. TAMM[2], A. O. TAMM[3], I. TRAAT[4]

[1] *Department of Statistics, University of Stockholm,*
Stockholm, SE-10691, Sweden
E-mail: Tatjana.vonRosen@stat.su.se

[2] *Department of Sports Medicine and Rehabilitation, University of Tartu,*
Tartu, 50406, Estonia
E-mail: ann.tamm@kliinikum.ee

[3] *Clinic of Internal Medicine, University of Tartu,*
Tartu, 50406, Estonia
E-mail: agu.tamm@kliinikum.ee

[4] *Institute of Mathematical Statistics, University of Tartu,*
Tartu, 50409, Estonia
E-mail: imbi.traat@ut.ee

Two basic components of the radiographically measured knee osteoarthritis (KOA), joint space width (JSW) and size of osteophytes (OPH), were studied in order to characterize early stage of KOA and to reveal additional aspects of grading it. The study was conducted in 161 individuals (100 women and 61 men, 34–54 years old). A linear mixed and a generalized linear models were used to identify risk factors for JSW and OPH. Accounting for intra- and inter-knee correlations in these models is novel for knee studies.

We obtained that at the early stage of KOA, JSW is more constitution-related, while the appearance of OPH more disease-related. The effect of gender on JSW was sufficiently large to establish different definitions of KOA for genders. In this case attention should be paid to the medial compartment as the narrower one. Our findings also suggested that the height of women could be taken into account when grading KOA, and that the imbalance between knee and thigh circumferences might refer to the presence of OPH.

Keywords: Correlated measurements; Generalized linear model; Joint space width; Linear mixed model; Osteophyte.

1. Introduction

Knee osteoarthritis (KOA) is known to cause complaints, decrease in working capacity and development of disability.[1] Among the recognized risk factors for KOA are aging, heredity, gender, high Body Mass Index (BMI, weight/height^2) and knee trauma.[2] For diagnosis of the KOA, besides clinical criteria, standard radiographs have been accepted as the most important technique.[1] At the level of the knee joint, and in particular of the femorotibial compartment, the recommended standard radiographic protocol has been the standing antero-posterior fully extended knee view.

The presence of osteophytes (OPH) and the narrowing of the joint space width (JSW) are the two main characteristics on which the radiographic diagnosis is based.[3,4] The latter is expressed as an osteoarthritis (OA) grade. OA grade as a summary measure expresses much less information about the stage of the disease than its components separately. Thus, in our study (likewise Refs. 5–10), we directly approached to the components of radiographic OA (JSW and OPH score), and looked for the factors affecting them in order to better understand the early diagnostics of KOA.

Most of the KOA studies have been addressed only to the tibiofemoral (TF) but not to the patellofemoral (PF) joint. Nevertheless, it has appeared that OA can occur more frequently in PF joint,[1,10,11] and it is similarly associated with development of disability.[12] The relative contribution of exogenous and constitution-related factors in the KOA of the TF and PF compartments might vary.[13] Our study covers both of these joints in both knees.

There are few data about the early stage of the KOA process in young and middle-aged population. Usually elderly people with severe disease are studied.[5] Our target group is a 34–54 years old population with symptomatic knee OA (pain, stiffness, difficulties in movement). People without knee symptoms matched by age and gender were used as controls for some statements.

Osteoarthritic studies have often faced the difficulty how to handle so called paired organ problem, for example, two knees and one person problem.[14] Simple solutions like choosing only one knee (random, fixed/sicker) or averaging measurements over knee yield the loss of information,[14] whereas direct incorporation of the information about both knees into standard statistical analysis lacks the assumption of independence and may give incorrect results. We shall use appropriate statistical methodology which is able to handle this problem by accounting for the intra- and inter-knee correlations.

The aim of the paper was to find out additional aspects of grading and diagnostics of osteoarthritis, by studying associations between KOA symptoms and JSW on one side, and OPH on another side. The effects of the patient related covariates as age, sex, height, weight, and knee related covariates as, for example, knee and thigh circumferences, knee (left, right), joint (PF, TF) and joint compartment (lateral, medial) were investigated. It is worth mentioning that in the recent radiographic atlas[15] of grading KOA gender differences were already accepted.

2. Subject selection

The study was conducted in Elva, a small town in Southern Estonia. The total population of Elva at the time of the study was 6292 persons served by four family doctors (FD). Persons aged 34–54 years, altogether 1916 individuals (891 male, 1025 females), formed 30.5% of the total population of the town. One of the four FD centers was chosen for the study, and the primary postal questionnaire was sent to all 559 individuals (274 males, 285 females) of age 34–54 served by this center. Questions related to the symptomatic KOA (Have you had periods of pain which lasted longer than one month, morning stiffness up to 30 min, crepitus or difficulties in movement in the knees, earlier knee trauma, or difficulties during certain activities (jogging, kneeling, jumping, climbing the stair)), were asked. Altogether 348 individuals (154 males, 194 females) responded (see Figure 1). Out of them 220 persons responded positively to the knee-related questions, and 161 (73%) person agreed to attend the in-depth clinical examinations. These people, 100 females (63%) and 61 males (37%), formed the symptomatic study group with average age 45 (SD 6) years. The average BMI was 28.2 (SD 5.8) kg/m^2 for females and 26.7 (SD 3.7) kg/m^2 for males. The 68 out of 128 persons who responded negatively to the knee-related questions, i.e. non-symptomatic subjects (28 males and 40 females), were invited to the in-depth examinations as controls. Their average age was 44.9 (SD 5.1) years.

In addition to sex, age, height and weight, the circumferences of both knees and thighs of participants were recorded. The height of a person was measured with accuracy 0.2 cm and weight was measured with accuracy 0.2 kg. Additionally, BMI (kg/m^2) was calculated. The knee and thigh (in lower 1/3 of thigh) circumferences were measured in sitting position, legs in 90° flexion with accuracy 0.2 cm. Difference between knee and thigh circumference (circdiff) was calculated in both legs, and their difference was found over legs (diff of circdiff).

The Ethical Committee of the Faculty of Medicine of the University of Tartu granted approval for this study.

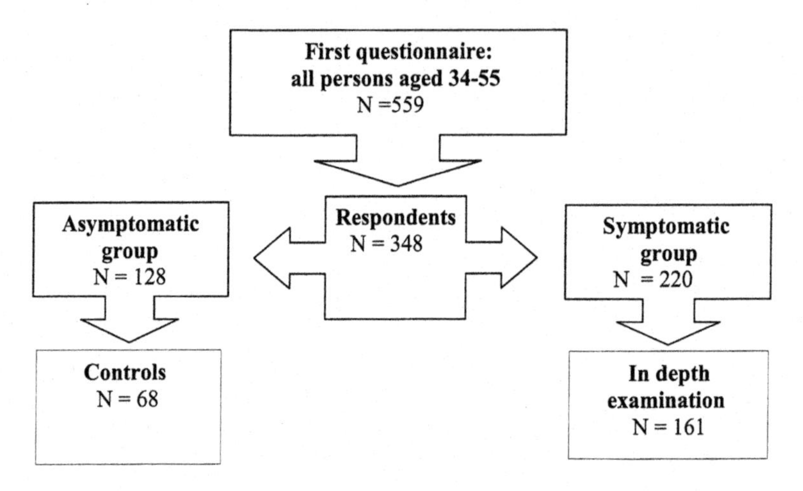

Fig. 1. Design of the Elva study population.

3. Measurements

Standard anteroposterior (AP) radiographs of TF joint were taken in a standing position (weight bearing, full extension) and axial radiographs of PF joint with prone technique, knee flexion $> 90°$. Each knee was radiographed separately. Two independent radiologists, unaware of the patients' complaints as well as the results of the functional performance and laboratory tests, read the radiographs. The first radiologist measured minimal JSW using a standard transparent ruler. The second radiologist digitally measured minimal JSW to the nearest 0.1 mm after scanning the radiographs. The medial and lateral JSW of the TF and PF compartments of both knees were measured. Osteophyte sizes (grades 0-3) were recorded in the lateral and medial compartments and classified according to Nagaosa et al.[15] When the first gradings were discordant, the final score was fixed by consensus agreement between the same two radiologists during the second assessment. With respect to the OA grade our study population was distributed as follows: grade 0 – 34%, grade I – 55%, grade II – 9% and grade III – 2%.

4. Statistical analysis

First, elementary statistical tools such as descriptive statistics and Pearson correlations were used to describe the study subjects by their various characteristics such as age, sex, height, weight, BMI, JSW and presence of OPH in different compartments of the knee. The OPH, initially measured using four-level scale (grades 0–3), were transformed into binary variable OPH-B with value 1 meaning presence of osteophyte (grade 1–3) and 0 meaning no osteophyte (grade 0). The positive correlation between binary variables says that probability is higher for simultaneous existence of OPH in two groups of interest.

To examine the influence of different factors such as age, sex, weight, height and knee compartment on the JSW, we used the following mixed linear model:[16,17]

$$\boldsymbol{y} = \boldsymbol{X\beta} + \boldsymbol{Z\gamma} + \boldsymbol{\varepsilon},$$

where $\boldsymbol{y} : 8n \times 1$ is the vector of observations (n individuals, each with 8 measurements), $\boldsymbol{\beta} : p \times 1$ is the vector of fixed effects (e.g. height, weight, age, gender, knee compartment), $\boldsymbol{X} : 8n \times p$ is a known design matrix for the fixed effects, $\boldsymbol{\gamma} : s \times 1$ is the vector comprising two random effects (individual, knee), $\boldsymbol{Z} : 8n \times s$ is a known incidence matrix for the random effects, $\boldsymbol{\varepsilon} : 8n \times 1$ is the vector of random errors. It is assumed that $\boldsymbol{\gamma}$ and $\boldsymbol{\varepsilon}$ are independently normally distributed random vectors with zero means and variances-covariance matrices $\boldsymbol{\Sigma}_{\gamma}$ and $\boldsymbol{\Sigma}_{\varepsilon} = \sigma^2 \boldsymbol{I}_{8n}$, respectively. Hence,

$$Var(\boldsymbol{y}) = \boldsymbol{Z\Sigma}_{\gamma}\boldsymbol{Z}' + \boldsymbol{\Sigma}_{\varepsilon} = \boldsymbol{I}_n \otimes \boldsymbol{\Sigma}.$$

The matrix $\boldsymbol{\Sigma}$ defines the dependency structure of the 8 JSW measurements within an individual and is as following:

$$\boldsymbol{\Sigma} = \boldsymbol{I}_2 \otimes \boldsymbol{\Sigma}_0 + (\boldsymbol{J}_2 - \boldsymbol{I}_2) \otimes \boldsymbol{\Sigma}_1,$$

where matrices $\boldsymbol{\Sigma}_0$ and $\boldsymbol{\Sigma}_1$ specify the intra-knee and inter-knee dependency structures of JSW measurements, respectively. The symbol $\otimes$ stands for the Kronecker product, $\boldsymbol{I}_m$ is the identity matrix of order m, $\boldsymbol{J}_m$ is a square matrix of size m with all elements equal to one.

Normality of the JSW was tested using the Shapiro-Wilk normality test.[18]

To study OPH we used a generalized linear model for binary response variable. The binary response variable was presence or absence of osteophytes (OPH-B) in knee locations, and the examined covariates were age, sex, height, weight, BMI, knee and thigh circumferences, also factors such as knee, joint and compartment. Since each subject contributed two knees with four OPH-B measurements on each, we handled them as correlated within-subject measurements with unstructured correlation matrix. We used generalized estimating equations[19] technique for our model offered by SAS 9.2 procedure GENMOD.

To test whether or not the JSW and OPH are related to pain and other knee symptoms we, respectively, used the independent two sample t-test and odds ratios while comparing symptomatic and non-symptomatic groups.

All the statistical analyses were performed using SAS system for Windows (Version 9.2, SAS Institute Inc., Cary, NC, USA). P-values, p less than or equal to 0.05 were considered to be statistically significant.

5. Results

An overview of the basic anthropometric characteristics of the study group, as well JSW and OPH-B tabulated by knee compartments, is given in Table 1.

Correlation analysis for JSW in different compartments of the knee shows (Table 2, 3) that we have highly positively correlated within-subject JSW data which confirm symmetry of a human body.

The highest correlations ρ, ranging between 0.7–0.8 occur between knee compartments symmetrically apart from the longitudinal axis in left and right sides, i.e. the correlations between the corresponding compartments in the left and right legs.

We found that males had significantly wider JSW than females ($p<0.0001$), JSW means difference is 0.73 (SD 0.27) mm. Since there was a strong gender effect on JSW, separate analyses were performed for men and women.

For both, males and females, a significant difference in JSW between the lateral and medial compartments of the knee was established: JSW at lateral compartment was significantly larger comparing to medial (Figure 2). JSW means difference for males is 0.67mm, 95% CI (0.49; 0.84), and for females 0.59mm, 95% CI (0.48; 0.72). Results also showed that the PF joint space was significantly narrower comparing to TF with means difference (PF minus TF) -0.60mm, 95% CI (-0.83; -0.38) for males, and -0.22mm, 95% CI (-0.33; -0.09) for females (Figure 2).

In modeling JSW no significant association with age in interval 34-54 years was seen ($p > 0.05$ for both, men and women). Our results showed that BMI did not contribute to JSW (for both men and women), as well

Table 1. Summary characteristics of the study group.

	Males (mean, SD)		Females (mean, SD)	
No. of subjects	61		100	
Age	45.0	(6.4)	45.6	(5.9)
Height (cm)	178.3	(6.7)	164.5	(6.2)
Weight (kg)	84.8	(13.6)	76.5	(16.8)
BMI (kg/m2)	26.7	(3.7)	28.2	(5.8)
Diff of circdiff (cm)	0.1	(1.3)	0.1	(1.7)
Left knee	JSW (mm) mean (SD)	OPH-B count, odds	JSW (mm) mean (SD)	OPH-B count, odds
lateral PF	5.8 (1.0)	11, 0.2	5.4 (0.9)	7, 0.1
medial PF	5.1 (1.6)	9, 0.2	4.6 (1.4)	16, 0.2
lateral TF	6.4 (1.0)	3, 0.1	5.5 (0.9)	7, 0.1
medial TF	5.9 (0.8)	2, 0.0	4.9 (0.8)	17, 0.2
circdiff (cm)	-4.0 (2.2)		-6.2 (3.1)	
Right knee	JSW (mm) mean (SD)	OPH-B count, odds	JSW (mm) mean (SD)	OPH-B count, odds
lateral PF	6.0 (1.0)	9, 0.2	5.4 (1.0)	7, 0.1
medial PF	5.2 (1.8)	5, 0.1	4.6 (1.5)	9, 0.1
lateral TF	6.5 (1.0)	4, 0.1	5.4 (0.8)	3, 0.0
medial TF	5.8 (0.8)	9, 0.2	5.1 (0.9)	19, 0.2
circdiff (cm)	-3.9 (2.2)		-6.1 (2.9)	

Note: JSW is characterized by the mean and the standard deviation (SD), osteophytes (OPH) by the count and odds in each locations where the measurements were made. The count shows the number of people having OPH in the considered location, the odds are calculated as a ratio of this count to the number of people without OPH in that location.

Table 2. Partial height-adjusted Pearson correlation coefficients of JSW in eight knee locations (males).

	PF L/L	PF L/M	TF L/L	TF L/M	PF R/L	PF R/M	TF R/L	TF R/M
PF/L/L	1.00							
PF/L/M	0.34	1.00						
TF/L/L	0.25	0.12	1.00					
TF/L/M	0.21	0.29	0.28	1.00				
PF/R/L	**0.75**	0.36	0.23	0.34	1.00			
PF/R/M	0.30	**0.79**	0.19	0.27	0.33	1.00		
TF/R/L	0.23	0.17	**0.78**	0.30	0.21	0.13	1.00	
TF/R/M	0.15	0.17	0.35	**0.74**	0.31	0.23	0.24	1.00

Note: Joint/Knee/Compartment; joint (TF – tibiofemoral, PF – patellofemoaral), knee (L – left, R – right), compartment (M – medial, L – lateral).

Table 3. Partial height-adjusted Pearson correlation coefficients of JSW in eight knee locations (females).

	PF L/L	PF L/M	TF L/L	TF L/M	PF R/L	PF R/M	TF R/L	TF R/M
PF/L/L	1.00							
PF/L/M	0.35	1.00						
TF/L/L	0.29	0.21	1.00					
TF/L/M	0.27	0.48	0.36	1.00				
PF/R/L	**0.84**	0.32	0.34	0.31	1.00			
PF/R/M	0.40	**0.84**	0.19	0.32	0.34	1.00		
TF/R/L	0.22	0.18	**0.77**	0.35	0.21	0.18	1.00	
TF/R/M	0.21	0.39	0.47	**0.79**	0.20	0.29	0.33	1.00

Note: Joint/Knee/Compartment; joint (TF – tibiofemoral, PF – patellofemoaral), knee (L – left, R – right), compartment (M – medial, L – lateral).

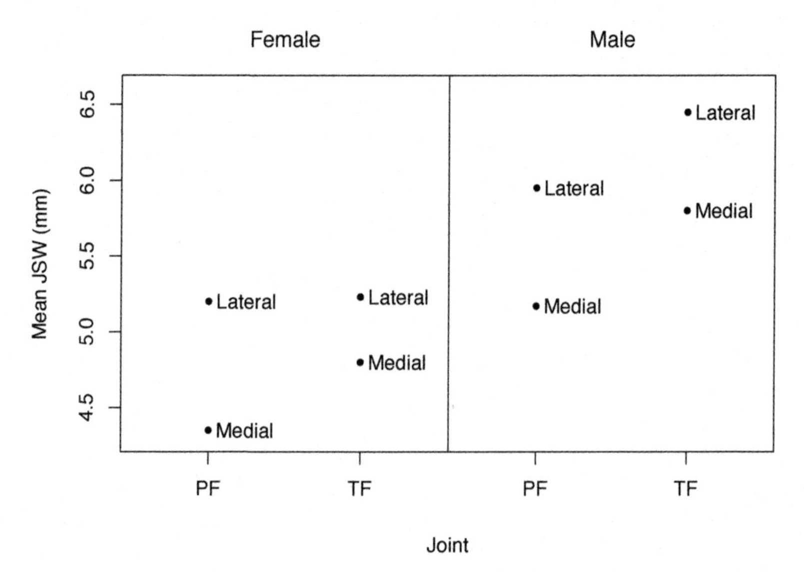

Fig. 2. Mean Joint Space Width adjusted for gender, joint and joint compartment. Note differences between genders, between TF/PF joints and between medial/lateral compartments.

as body weight itself (see Table 4). Neither knee circumferences were significantly associated with JSW ($p = 0.9$ and $p = 0.8$ for men and women, respectively).

In men group the effect of height on JSW was weak (regression coefficient 0.006, $p = 0.6996$), but in women group there was a strong positive

Table 4. Factors associated with Joint Space Width.

	$\hat{\beta}$	95% CI	p-value
		males	
Height	0.006	(-0.023; 0.034)	0.6996
Weight	0.0004	(-0.015; 0.016)	0.9627
Age	0.007	(-0.024; 0.038)	0.6367
Joint (PF vs TF)	-0.604	(-0.826; -0.381)	< 0.0001
Compartment (L vs M)	0.670	(0.492; 0.841)	< 0.0001
OPH (0 vs 1)	0.101	(-0.223; 0.425)	1.0000
OPH (0 vs 2)	0.145	(-1.071; 1.360)	1.0000
		females	
Height	0.050	(0.029; 0.072)	< 0.0001
Weight	0.006	(-0.002; 0.015)	0.1483
Age	-0.018	(-0.041; 0.004)	0.1046
Joint (PF vs TF)	-0.216	(-0.332; -0.099)	0.0003
Compartment (L vs M)	0.597	(0.480; 0.715)	< 0.0001
OPH (0 vs 1)	0.064	(-0.185; 0.312)	1.0000
OPH (0 vs 2)	0.958	(0.320; 1.597)	0.0010

Note: OPH – osteophytes; PF – patellofemoral, TF – tibiofemoral; L–lateral; M – medial.

association between JSW and height (regression coefficient 0.05, $p < 0.0001$). Women 10 cm taller have on the average 0.5 mm large JSW.

The grade of OPH (0–3) was also used as an explanatory variable for JSW. Differently from males, the effect of the grade of OPH for women was statistically significant (p<0.01): higher grade of OPH was associated with a reduction in JSW, for example women with grade 2 of OPH have 0.96 mm decrease in JSW comparing to women without OPH. There was no statistical difference in JSW between women without OPH (grade=0) and those who have small OPH (grade=1). In addition, the two-way interaction effect (Figure 2) of the compartment (lateral or medial) and the joint (PF or TF) was statistically significant ($p = 0.015$), i.e. for females the changes in JSW in lateral and medial compartments of PF and TF joints are not homogeneous.

To evaluate association between knee symptoms and the JSW, and the OPH, symptomatic and non-symptomatic groups were compared (independent two-sample t-tests and odds-ratios). For both males and females, there was neither strong association between JSW and symptoms, nor significant difference between knees with respect to the average JSW ($p > 0.05$). Opposite to JSW the OPH were strongly associated with knee symptoms. The odds ratio based on an *subject-related* summary measure SUM-OPH (1 – person has OPH, 0 – does not have) revealed that knee symptoms increase the odds of OPH 3.55 times (OR=3.55, 95% CI=(1.73; 7.29)). Furthermore, based on a *joint-related* summary measure (1 – OPH in any location of a joint, 0 – no OPH in that joint), the knee complaints increased the odds of OPH in both the TF and PF joints, the ORs were 3.93, 95% CI (1.48; 10.48) and 2.54, 95% CI (1.16; 5.56), respectively.

The factors affecting OPH were analyzed more thoroughly in the group having knee symptoms. Differently from JSW case gender effect to OPH was not significant (OR=1.36, 95% CI=(0.70, 2.64)). The effect appears through interactions, as we see later.

The correlation matrix for OPH-B (men women jointly) showed pattern similar to the JSW case (Tables 2, 3), revealing that OPHs tend to be simultaneously in the knee compartments symmetrically apart from the longitudinal axis of a person, this especially for the PF joint. The correlation was 0.80 between lateral compartments of PF joint (left versus right knee) and 0.72 between corresponding medial compartments; 0.58 between lateral compartments of TF joint and 0.52 between corresponding medial compartments. Correlations between other locations were weak positive (around 0.2–0.3) or zero. Symmetricity, right versus left, was also confirmed by McNemar's test for paired samples: no significant difference with respect to OPH over knees ($p>0.05$) was discovered. When restricting comparison to radiologically advanced cases (grade II and III), we noticed that in PF joint both knees were symmetrically affected in 16 cases out of 17, and in TF joint OA features were found in all 7 cases in both knees.

Table 5. Odds ratios for osteophytes in knees.

Effect name	OR	95% CI	p-value
Height	0.94	(0.89; 0.99)	0.017
Weight*F	1.07	(1.05; 1.10)	<0.001
Weight*M	1.04	(1.00; 1.07)	0.065
JSW*F	0.73	(0.56; 0.94)	0.013
JSW*M	1.02	(0.80; 1.31)	0.858
Circdiff	1.20	(1.08; 1.33)	0.001
Diff of circdiff	1.28	(1.05; 1.56)	0.014

Note: M – male, F – female; Circdiff – difference between knee and thigh circumference; Diff of circdiff – difference in circdiff between left and right legs.

A generalized linear model was fitted to OPH-B having within-subject correlations due to measurements from four locations in both knees. The resulting effects of continuous explanatory variables are given in Table 5.

The effect of factors such as knee, joint and compartment was also examined in the model. Neither knee nor joint did affect odds of OPH ($p > 0.05$), i.e. OPH were equally likely in right versus left knee, as well in TF versus PF joint. Their effect revealed through interaction with compartment and sex. For females the odds of OPH were significantly higher ($p < 0.05$) in medial compartments for both right and left TF, and left PF (Figure 3 for logarithms of odds).

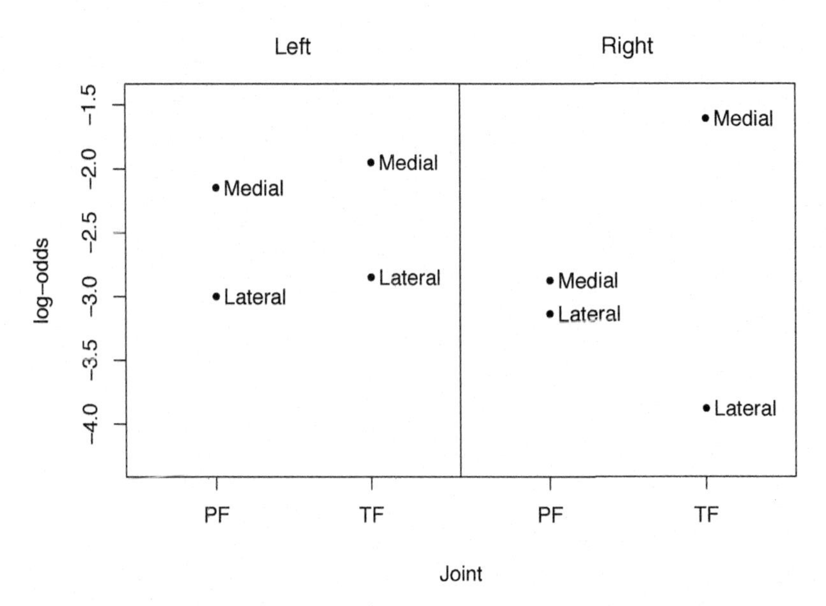

Fig. 3. Logarithms of odds of osteophytes for females by knee, joint and joint compartment (lateral versus medial). Note differences between right and left knee, medial and lateral compartments.

Table 5 shows that the height of a person had decreasing (OR 0.94, $p = 0.017$) whereas the weight had increasing effect to the odds of having OPH. However, different increasing effects by genders (OR 1.07 for females and 1.04 for males) is insufficient evidence for concluding that weight effect in females is bigger than that in males. The JSW was used as an explanatory variable for OPH. Its effect was seen for females; the larger was JSW the smaller were odds of OPH. Finally, there were significant associations between OPH and covariates expressing imbalance of knee and thigh circumferences ($p = 0.001$, $p = 0.014$). The bigger was knee circumference compared to the thigh circumference, the higher were odds of OPH. In addition, the increase of this difference between the right and left legs (circdiff) increased the odds of having OPH. Remarkable was the absence of the age effect ($p > 0.05$).

6. Discussion

The aim of the paper was to model JSW and presence of OPH in young and middle-aged population by examining the effects of the person-related covariates as age, sex, height, weight, and knee-related covariates/factors such as knee (left, right), joint (PF, TF) and knee compartment (lateral, medial), knee-thigh circumferences. The sex- and age-matched controls were used to examine the effect of knee complaints to JSW and OPH. We used contemporary statistical methods (a linear mixed model and a generalized linear model with estimating equations) that take into account intra- and inter-knee correlations among the JSW and OPH measurements. The challenge was to investigate an early stage of the KOA process and to differentiate effects on the JSW and OPH as being constitution-related or disease-related. For example, among the recognized risk factors for KOA, such as heredity, gender, high BMI, aging and knee trauma,[2] the first three might express a constitution-related influence on joints.

As reported in Refs. 20 and 21, the JSW does not decrease with age of healthy people (without knee pain and OPH) in age group 40–80. Our study results confirmed these findings for our symptomatic group aged 34–54. In that group (adjusted for height) age did not affect formation of OPH, although one can expect positive association between age and the size of OPH in older people.[22]

Although BMI is an established risk factor for KOA, we found that BMI, as well as weight itself, do not contribute significantly to the JSW at all sites of the knee (medial and lateral sites of PF and TF joints) in both males and females. This is in line with results obtained for healthy people[20] and for people with knee pain (42–77 years).[22] Contrary to[20] which reports no significant correlation between JSW and height in healthy people, we have found a strong positive height effect on JSW in females.

The positive correlation between BMI and the size of outward OPH, and the presence of a negative association between JSW and OPH was reported in Ozdemir et al.[22] Our findings also confirm that the weight has significant positive effect on odds of having OPH. Opposite to the weight, the height has decreasing effect on OPH. Thus, tall people, especially women, seem to be in a good situation with respect to the radiographic KOA: larger JSW and smaller odds of having OPH.

JSW depends significantly on sex: males have significantly wider JSW than females.[20,24] At the same time the odds of OPH do not differ significantly between males and females. The effect of sex on OPH can be seen through interactions. The interaction with knee compartment reveals that females have OPH more frequently on the medial compartment, which in this case can indicate that for men and women changes in the knee are not homogeneous.

Neame et al.[24] noted some left-right asymmetry in TF joint studing older people than we did (mean age 63.7). In our study, left versus right tests have confirmed no significant difference between knees neither with respect to the average JSW, nor OPH, for both males and females. This confirmed symmetric nature of a human body, i.e lateral compartments of joints tend to be similar in both legs, the same holds for medial compartments.

Comparing lateral versus medial compartments, a significant difference in JSW can be seen, joint space at lateral side is wider (see Figure 2). This matches Wolfe et al.[25] who have seen more decrease of JSW in medial compartment than lateral. Comparing TF versus PF, one sees significantly wider JSW in the TF joint (see Figure 2).

In case of symptomatic knee osteoarthritis it was established that presence of OPH, but not JSW narrowing, is associated with pain.[20] This is confirmed by our study. Moreover, our results suggest that in OA progression OPH appear earlier than the JSW narrowing. Presence of the knee symptoms yields the increase in the odds of having OPH, OR=3.55. Furthermore, higher odds of having OPH appears in both joints. At the same time JSW does not differ significantly among the people with and without symptomatic KOA, this is valid for both males and females.

For symptomatic group the increase of knee circumference (relative to thigh circumference) increases the odds of OPH, OR=1.2. The difference of this measure over knees (diff of circdiff) is also associated with OPH in knees. At the same time JSW does not depend neither on the knee not thigh circumferences.

Our opinion is that JSW is both a constitutional indicator of a human body as well linked to OA. Particularly in females it was strongly related to the height of a person. This implies that narrower JSW is not always associated with the disease. Looking for the factors that influence JSW we therefore considered different measurements of a human body.

In the contrast to JSW, the OPH are more directly disease related. They are not originally present in the knee of the healthy person and develop later. They are significantly associated with cartilage defects.[26] We therefore looked for the characteristics expressing the imbalance of knee measurements, for example the one used in this study is the difference between thigh and knee circumferences.

Further research interests concern the beginning stage of the KOA, in particular, the formation of osteophytes and the joint space narrowing. We planned the longitudinal study in order to understand better the process of JSN, for example speed and homogeneity over different knee compartments, and investigate the possibility to take into account the changes in JSW when diagnosing/grading KOA.

7. Conclusions

In our study, both JSW and OPH of respective joint compartments in right and left knees exhibited strong positive correlations (symmetric nature of a human body). As a rule, medial JSW was narrower than lateral JSW both in the TF and PF joints of both genders. A strong gender effect on JSW (0.73 mm wider for males) was observed, whereas OPH were not affected by gender. Height had an increasing effect for female JSW and a decreasing effect for odds of OPH of both genders (OR 0.94). Contrary to JSW, OPH were strongly associated with knee complaints (OR 3.55), with weight (OR 1.07 for females and 1.04 for males) and with imbalance in knee and thigh circumferences. Narrowing of JSW was associated with presence of OPH.

We conclude that at the early stage of KOA, JSW should be considered more constitution-related, while the appearance of OPH more disease-related. The effect of gender on JSW was sufficiently large to justify different definitions of KOA for genders. In this case attention should be paid to the medial compartment as the narrower one. Our findings also suggested that the height of women could be taken into account when grading OA, and that the imbalance between knee and thigh circumferences refers to the presence of OPH.

Acknowledgments

We thank Mare Lintrop and Karin Veske for their valuable contribution in reading radiographs.

The work was supported by the Swedish Research Council (T. von Rosen by the grant 2007-890) and by the Estonian Science Foundation (A. E. Tamm, A. O. Tamm by the grant 5308, and I. Traat by the grant 8789 and by Targeted Financing Project SF0180015s12).

References

1. D. T. Felson and Y. Zhang, An update on the epidemiology of knee and hip osteoartritis with view to prevention. *Arth. Rheum* **41**, 1343–55 (1998).
2. F. M. Cicuttini, T. Spector and J. Baker, Risk factors for osteoarthritis in the tibiofemoral and patellofemoral joints of the knee. *J. Rheumatol.* **24**, 1164–7 (1997).
3. J. C. Buckland-Wright and J. Scott, C. Peterfy, Radiographic imaging techniques. *Osteoarthritis and Cartilage* **4**, 238–240 (1996).
4. J. C. Buckland-Wright, Workshop for concensus in OA imaging. Review of the anatomical and radiological differences between fluoroscopic and non-fluoroscopic positioning of osteoarthritic knees. *Osteoarthritis and Cartilage* **14**, A19–A31 (2006).
5. P. Creamer, M. Lethbridge-Cejku and M. C. Hochberg, Factors associated with functional impairment in symptomatic knee osteoarthritis. *Rheumatology* **39**, 490–496 (2000).
6. P. S. Emrani, J. N. Katz, C. L. Kessler et al., Joint space narrowing and Kellgren–Lawrence progression in knee osteoarthritis. *Osteoarthritis Cartilage* **16**(8), 873–882 (2008).
7. T. E. McAlindon, S. Snow, C. Cooper and P. A. Dieppe, Radiographic patterns of osteoarthritis of the knee joint in the community: importance of the patellofemoral joint. *Ann Rheum Dis* **51**, 844–9 (1992).
8. V. K. Srikanth, J. L. Fryer, G. Zhai and T. M. Winzenberg, A meta-analysis of sex differences: prevalence, incidence and severity of osteoarthritis. *Osteoarthritis Cartilage* **13**, 769–81 (2005).
9. S. L. Zeger, K. Y. Liang and P. Albert, Models for longitudinal data, a generalized estimating equation approach. *Biometrics* **44**, 1049–1060 (1988).
10. S. A. Mazzuca, K. D. Brandt, N. C. German, K. A. Buckwalter, K. A. Lane and B. P. Katz, Development of radiographic changes of osteoartritis in the Chingford knee. *Ann Rheum Dis* **62**, 1061–1065 (2003).
11. A. Tamm, M. Lintrop, K. Veske, U. Hansen and A. Tamm, Prevalence of patello- and tibiofemoral osteoarthritis in Elva, southern Estonia. *Journal of Rheumatology* **35**(3), 543–544 (2008).
12. K. D. Brandt, M. Doherty and S. L. Lohmander, Introduction: the concept of osteoarthritis as failure of the diarthrodial joint, in *Osteoarthritis*, 2nd edn., eds. K. D Brandt, M. Doherty, S. L. Lohmander, pp. 69–71 (Oxford 2003).
13. C. Cooper, T. McAlindon, S. Snow, K. Vines, P. Young, J. Kirwan and P. Dieppe, Mechanical and constitutional risk factors for symptomatic knee osteoarthritis: differences between medial tibiofemoral and patellofemoral disease. *J. Rheumatol.* **21**, 307–13 (1994).
14. A. J. Sutton, K. R. Muir and A. C. Jones, Two knees or one person: data analysis strategies for paired joints or organs. *Ann Rheum Dis.* **56**, 401–402 (1997).
15. Y. Nagaosa, M. Mateus, B. Hassan, P. Lanyon, and M. Doherty, Development of logically devised line drawing atlas for grading of knee osteoarthritis. *Ann Rheum Dis* **59**, 587–595 (2000).
16. S. R. Searle, *Linear Models* (John Wiley & Sons, Inc, New York, 1971).

17. E. Demidenko, *Mixed Models: Theory and Applications* (Wiley, New York, 2004).
18. S. S. Shapiro, M. B. Wilk, An analysis of variance test for normality (complete samples). *Biometrika* **52**, 591–611 (1965).
19. J. W. Hardin and J. M. Hilbe, *Generalized Estimating Equations* (Chapman & Hall/CRC, London, 2003).
20. P. Lanyon, S. O'Reilly, A. Jones and M. Doherty, Radiographic assessment of symptomatic knee osteoarthritis in the community: definition and normal joint space. *Ann Rheumatic Diseases* **57**(10), 595–601 (1998).
21. K. A. Beattie, P. Boulos, J. Duryea and M. Pui et al., Normal values of medial tibiofemoral minimum joint space width in the knees of healthy men and women. *Osteoarthritis and Cartilage* **12**, B, 965–66 (2004).
22. F. Ozdemir, O. Tukenmez, S. Kokino and F. N. Turan, How do marginal osteophytes, joint space narrowing and range of motion affect each other in patients with knee osteoarthritis. *Rheumatol Int*, DOI 10.1007/s00296-005-0016-0 (2005).
23. J. E. Dacre, D. L. Scott, J. A. Da Silva,G. Welsh and E. C. Huskisson, Joint space in radiologically normal knees. *Br J Rheumatol* **30**(6), 426–8, (1991).
24. R. Neame, W. Zhang, C. Deighton, M. Doherty, S. Doherty, P. Lanyon and G. Wright, Distribution of radiographic osteoartritis between the right and left hands, hips, and knees. *Arthritis and Reumatism* **50**, 1487–1494 (2004).
25. F. Wolfe and N. E.Lane, The longterm outcome of osteoarthritis: rates and predictors of joint space narrowing in symptomatic patients with knee osteoarthritis. *J. Rheumatol* **29**, 139–46 (2002).
26. C. Ding, P. Garnero, F. Cicuttini, F. Scott, H. Cooley, and G. Jones, Knee cartilage defects: association with early radiographic osteoarthritis, decreased cartilage volume, increased joint surface area and type II collagen breakdown. *Osteoarthritis and Cartilage* **13**, 198–205 (2005).

SIMULTANEOUS CONFIDENCE REGION FOR ρ AND σ^2 IN A MULTIVARIATE LINEAR MODEL WITH UNIFORM CORRELATION STRUCTURE

I. ŽEŽULA* and D. KLEIN

Institute of Mathematics, P. J. Šafárik University,
Jesenná 5, SK-04154 Košice, Slovakia
**E-mail: ivan.zezula@upjs.sk*
www.upjs.sk

The article derives joint distribution of estimators of ρ and σ^2 in the standard GCM with uniform correlation structure and studies various possibilities of forming confidence regions for these estimators.

Keywords: Growth curve model, orthogonal decomposition, uniform correlation structure

1. Estimators and their distribution

The basic model that we consider is the following one:

$$\begin{aligned} Y &= XBZ' + \mathbf{e}, \\ \mathrm{vec}(\mathbf{e}) &\sim N\left(0, \Sigma \otimes I_n\right), \quad \Sigma = \sigma^2\left[(1-\rho)I_p + \rho\mathbf{1}\mathbf{1}'\right], \end{aligned} \tag{1}$$

where $Y_{n\times p}$ is a matrix of independent p-variate observations, $X_{n\times m}$ is an ANOVA design matrix, $Z_{p\times r}$ is a regression variables matrix, $\mathbf{e}$ is a matrix of random errors, and $\mathbf{1}$ is p-vector of 1's. As for the unknown parameters, $B_{m\times r}$ is an location parameters matrix, and σ^2, ρ are (scalar) variance and correlation parameters. The vec operator stacks elements of a matrix into a vector column-wise. Notice that throughout the article we suppose normal distribution of Y.

Assumed correlation structure is called uniform correlation structure or intraclass correlation structure. General problem of variance components estimation in the growth curve model (GCM) was studied by Žežula.[1] Special variance structures including uniform correlation structure were studied in the context of GCM first by Lee,[2] and later by Kanda,[3] Kanda,[4] Rao Chaganty,[5] Žežula,[6] Wu,[7] Wu,[8] Klein and Žežula,[9] Ye and Wang,[10] and recently by Žežula and Klein.[11]

Since we do not use matrix Z for the estimation of variance matrix Σ (see later), all the results of this article can also be applied to multivariate regression model

$$\begin{gathered} Y = XB + \mathbf{e}, \\ \operatorname{vec}(\mathbf{e}) \sim N(0, \Sigma \otimes I_n), \quad \Sigma = \sigma^2 [(1-\rho)I_p + \rho \mathbf{1}\mathbf{1}']. \end{gathered} \tag{2}$$

Let P_F be the orthogonal projection matrix onto the column space $\mathcal{R}(F)$ of a matrix F, and $M_F = I - P_F$ onto its orthogonal complement. Notice that we make use of projector $P_{\mathbf{1}}$ on the space of multiples of $\mathbf{1}$.

It is well known that $S = \frac{1}{n-r(X)} Y' M_X Y$ is the best estimator of Σ in the case of completely unknown structure under normality (here $r(X)$ is the rank of X). This estimator is a logical base for the variance parameters estimators in the case of a known structure. Then, simple estimators (see Žežula[6]) are

$$\hat{\sigma}^2 = \frac{\operatorname{Tr}(S)}{p}, \quad \hat{\rho} = \frac{1}{p-1}\left(\frac{\mathbf{1}'S\mathbf{1}}{\operatorname{Tr}(S)} - 1\right). \tag{3}$$

They can be expressed in the form (see Ye and Wang,[10] Žežula and Klein[11]):

$$\hat{\sigma}^2 = \frac{\operatorname{Tr}(V_1) + \operatorname{Tr}(V_2)}{p}, \quad \hat{\rho} = 1 - \frac{p \operatorname{Tr}(V_2)}{(p-1)(\operatorname{Tr}(V_1) + \operatorname{Tr}(V_2))}, \tag{4}$$

where

$$V_1 = P_{\mathbf{1}} S P_{\mathbf{1}}, \quad V_2 = M_{\mathbf{1}} S M_{\mathbf{1}}.$$

Since $P_{\mathbf{1}}$ and $M_{\mathbf{1}}$ are mutually orthogonal, it was shown in Žežula and Klein[11] that the distributions of the $\operatorname{Tr}(V_1)$ and $\operatorname{Tr}(V_2)$ are independent,

$$\begin{aligned} \operatorname{Tr}(V_1) &\sim \frac{\sigma^2[1+(p-1)\rho]}{n-r(X)} \chi^2_{n-r(X)}, \\ \operatorname{Tr}(V_2) &\sim \frac{\sigma^2(1-\rho)}{n-r(X)} \chi^2_{(p-1)(n-r(X))}, \end{aligned}$$

implying

$$\begin{gathered} \hat{\sigma}^2 \sim \frac{\sigma^2}{p(n-r(X))}\left[(1+(p-1)\rho)\chi^2_{n-r(X)} + (1-\rho)\chi^2_{(p-1)(n-r(X))}\right], \\ \frac{1-\rho}{1+(p-1)\rho}\left[\frac{1+(p-1)\hat{\rho}}{1-\hat{\rho}}\right] \sim F_{n-r(X),(p-1)(n-r(X))}. \end{gathered}$$

Moreover, having two independent distributions at the beginning, it is easy to derive the joint distribution of the parameter estimators.

Lemma 1.1. *Joint density of* $(\hat{\sigma}^2, \hat{\rho})$ *is*

$$f(s,r) = \exp\left\{-\frac{s\,(n-r(X))}{2\sigma^2}\left[\frac{1+(p-1)r}{1+(p-1)\rho} + \frac{(p-1)\,(1-r)}{1-\rho}\right]\right\}$$
$$\times \frac{p(p-1)^{(p-1)(n-r(X))/2}(n-r(X))^{p(n-r(X))/2}}{2^{p(n-r(X))/2}\Gamma\left(\frac{n-r(X)}{2}\right)\Gamma\left(\frac{(p-1)(n-r(X))}{2}\right)} \cdot \frac{1}{s}\left(\frac{s}{\sigma^2}\right)^{p(n-r(X))/2}$$
$$\times \frac{1}{1+(p-1)r}\left(\frac{1+(p-1)r}{1+(p-1)\rho}\right)^{(n-r(X))/2}\frac{1}{1-r}\left(\frac{1-r}{1-\rho}\right)^{(p-1)(n-r(X))/2},$$

for $s > 0, r \in \left(-\frac{1}{p-1};1\right)$.

Proof. Marginal densities of $\operatorname{Tr}(V_1)$ and $\operatorname{Tr}(V_2)$ are simple to derive from corresponding χ^2 densities. Because of independence, their joint density is the product of the marginal ones. Then we use transformation

$$t: \begin{pmatrix}\operatorname{Tr}(V_1)\\ \operatorname{Tr}(V_2)\end{pmatrix} \longrightarrow \begin{pmatrix}\hat{\sigma}^2\\ \hat{\rho}\end{pmatrix} = \begin{pmatrix}\frac{\operatorname{Tr}(V_1)+\operatorname{Tr}(V_2)}{p}\\ 1-\frac{p\operatorname{Tr}(V_2)}{(p-1)(\operatorname{Tr}(V_1)+\operatorname{Tr}(V_2))}\end{pmatrix}$$

to obtain (after some tedious calculations) joint density of $\left(\hat{\sigma}^2, \hat{\rho}\right)$. □

This density is unimodal. Example of such density is in Figure 1 (3D-plot) and 2 (contour plot). Level contours are almost elliptical for ρ close to 0, and more and more crescent-like when ρ goes to its limits. The mode lies close to true value (σ^2, ρ).

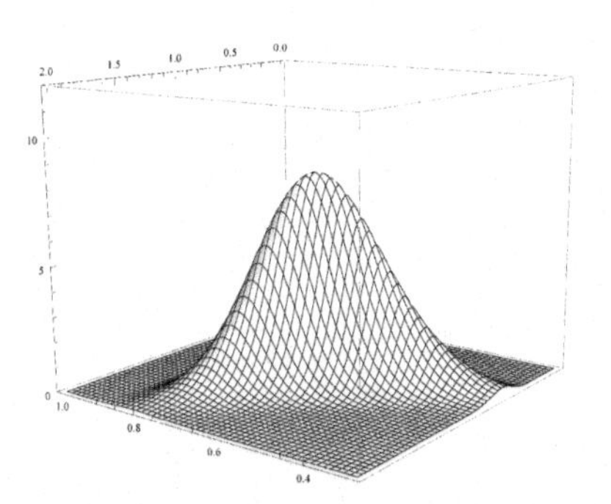

Fig. 1. Density 3D-plot.

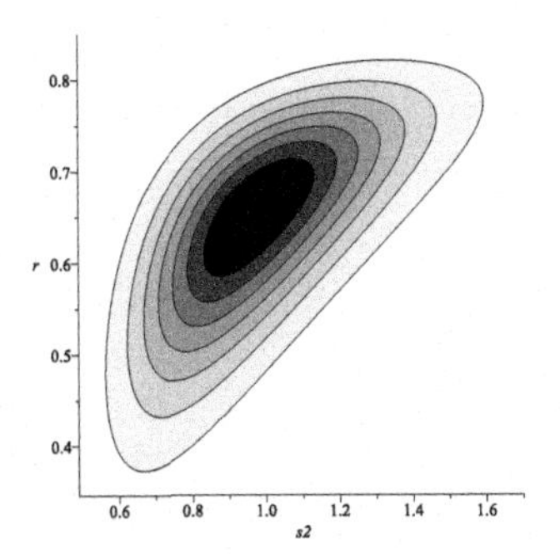

Fig. 2. Density contour plot.

2. Confidence regions

The one-dimensional distribution of $\hat{\rho}$ depends only on unknown value of ρ, and the function $\frac{1-\rho}{1+(p-1)\rho}$ is monotone in the region of possible values. Then, it is easy to provide $1-\alpha$ confidence interval for ρ. It is

$$\left(\frac{1-c_1}{1+(p-1)c_1}; \frac{1-c_2}{1+(p-1)c_2}\right), \tag{5}$$

where

$$c_1 = \frac{1-\hat{\rho}}{1+(p-1)\hat{\rho}} F_{n-r(X),(p-1)(n-r(X))}\left(1-\frac{\alpha}{2}\right)$$

and

$$c_2 = \frac{1-\hat{\rho}}{1+(p-1)\hat{\rho}} F_{n-r(X),(p-1)(n-r(X))}\left(\frac{\alpha}{2}\right).$$

The problem with $\hat{\sigma}^2$ is that its distribution depends on both σ^2 and ρ. The most primitive (and conservative) method is to look how does the variance of the estimator depend on ρ and take the least favourable value. A simple calculation yields that

$$\operatorname{var}\hat{\sigma}^2 = \frac{2\sigma^4\left(1+(p-1)\rho^2\right)}{p\left(n-r(X)\right)}.$$

Since it grows with ρ^2, we get the largest confidence interval for $\rho = 1$ (and the smallest one for $\rho = 0$). Figures 3 – 5 displays 0.025 and 0.975 quantiles of this distribution as a function of ρ for $\sigma^2 = 1$ and $n - r(X) = 20, 25, 50, 100$ for selected p's.

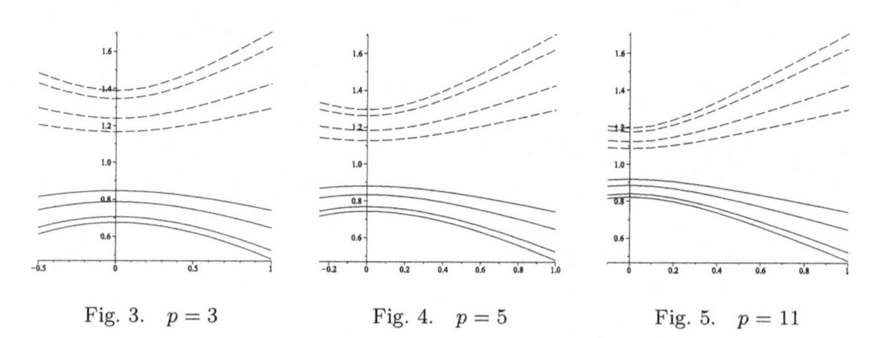

Fig. 3. $p = 3$ Fig. 4. $p = 5$ Fig. 5. $p = 11$

Simple computation gives $1-\alpha$ (conservative) confidence interval for σ^2 to be

$$\left(\frac{(n-r(X))\hat{\sigma}^2}{\chi^2_{n-r(X)}\left(1-\frac{\alpha}{2}\right)};\frac{(n-r(X))\hat{\sigma}^2}{\chi^2_{n-r(X)}\left(\frac{\alpha}{2}\right)}\right). \tag{6}$$

If it is reasonable to suppose that $\rho<0$, we can do better taking $\rho=\frac{-1}{p-1}$ as the least favourable value. This leads to the interval

$$\left(\frac{(p-1)(n-r(X))\hat{\sigma}^2}{\chi^2_{(p-1)(n-r(X))}\left(1-\frac{\alpha}{2}\right)};\frac{(p-1)(n-r(X))\hat{\sigma}^2}{\chi^2_{(p-1)(n-r(X))}\left(\frac{\alpha}{2}\right)}\right). \tag{7}$$

Because the variance for $\rho=1$ does not depend on p, the difference between $\rho=1$ and $\rho=0$ increases with p. Thus, especially for larger p it is better to look for simultaneous confidence region for pair (σ^2,ρ).

The smallest size confidence region must be based on α-cuts: if $\left(\hat{\sigma}_0^2,\hat{\rho}_0\right)$ are the observed values of estimators, $1-\alpha$ confidence region consists of all pairs (σ^2,ρ) for which $\left(\hat{\sigma}_0^2,\hat{\rho}_0\right)$ lies within contour $f_{(\sigma^2,\rho)}\left(\hat{\sigma}^2,\hat{\rho}\right)=c_\alpha$, where

$$\iint\limits_{f_{(\sigma^2,\rho)}(x,y)\geq c_\alpha} f_{(\sigma^2,\rho)}\left(x,y\right)\,dy\,dx=1-\alpha\,. \tag{8}$$

To find such a region is a computationally demanding task. To do that, we have to realize that p-value of the observed pair $\left(\hat{\sigma}_0^2,\hat{\rho}_0\right)$ with respect to true value $\left(\sigma^2,\rho\right)$ is

$$p=1-\iint\limits_{f_{(\sigma^2,\rho)}(x,y)\geq f_{(\sigma^2,\rho)}\left(\hat{\sigma}_0^2,\hat{\rho}_0\right)} f_{(\sigma^2,\rho)}\left(x,y\right)\,dy\,dx\,.$$

Then, the numerical algorithm can be the following one:

(1) Start with $\left(\sigma^2,\rho\right)=\left(\hat{\sigma}_0^2,\hat{\rho}_0\right)$. Choose step change $\Delta\sigma^2>0$.
(2) For current σ^2, find left and right boundaries for ρ, i.e. such values $\rho_l\left(\sigma^2\right)$, $\rho_r\left(\sigma^2\right)$ that p-value of $\left(\hat{\sigma}_0^2,\hat{\rho}_0\right)$ is equal to α. If such values do not exist, i.e. all p-values are less then α, go to (4).
(3) Exchange σ^2 for $\sigma^2+\Delta\sigma^2$, change starting point of ρ to the last $(\rho_l+\rho_r)/2$, and go to (2).
(4) After the first run, set $\sigma^2=\hat{\sigma}_0^2$, change the sign of $\Delta\sigma^2$, and go to (3). After the second run, stop.

Now we have a set of discrete points, convex hull of which (numerical solution of (8)) is approximate optimal confidence region for $\left(\sigma^2,\rho\right)$.

The previous algorithm takes days even with Mathematica 8 on a good 64-bit machine. That is why we have searched for something simpler, even if not optimal.

If we denote $\alpha^* = 1 - \sqrt{1-\alpha}$, then the following two inequalities hold with probability $1-\alpha$:

$$\begin{aligned}\frac{\sigma^2[1+(p-1)\rho]}{n-r(X)}\,\chi^2_{n-r(X)}\left(\tfrac{\alpha^*}{2}\right) &\le \operatorname{Tr}(V_1)\\ &\le \frac{\sigma^2[1+(p-1)\rho]}{n-r(X)}\,\chi^2_{n-r(X)}\left(1-\tfrac{\alpha^*}{2}\right)\\ \frac{\sigma^2(1-\rho)}{n-r(X)}\,\chi^2_{(p-1)(n-r(X))}\left(\tfrac{\alpha^*}{2}\right) &\le \operatorname{Tr}(V_2)\\ &\le \frac{\sigma^2(1-\rho)}{n-r(X)}\,\chi^2_{(p-1)(n-r(X))}\left(1-\tfrac{\alpha^*}{2}\right).\end{aligned}$$

Since $\operatorname{Tr}(V_1) = \hat{\sigma}^2\,[1+(p-1)\hat{\rho}]$ and $\operatorname{Tr}(V_2) = \hat{\sigma}^2(p-1)\,(1-\hat{\rho})$, this can easily be turned into

$$\begin{aligned}\frac{n-r(X)}{\chi^2_{n-r(X)}\left(1-\frac{\alpha^*}{2}\right)}\hat{\sigma}^2\,[1+(p-1)\hat{\rho}] &\le \sigma^2\,[1+(p-1)\rho]\\ &\le \frac{n-r(X)}{\chi^2_{n-r(X)}\left(\frac{\alpha^*}{2}\right)}\hat{\sigma}^2\,[1+(p-1)\hat{\rho}] \qquad (9)\end{aligned}$$

and

$$\begin{aligned}\frac{(p-1)(n-r(X))}{\chi^2_{(p-1)(n-r(X))}\left(1-\frac{\alpha^*}{2}\right)}\hat{\sigma}^2\,(1-\hat{\rho}) &\le \sigma^2\,(1-\rho)\\ &\le \frac{(p-1)(n-r(X))}{\chi^2_{(p-1)(n-r(X))}\left(\frac{\alpha^*}{2}\right)}\hat{\sigma}^2\,(1-\hat{\rho})\,. \qquad (10)\end{aligned}$$

If we take ρ as a function of σ^2, inequalities (9) defines two decreasing bounds for ρ for every σ^2:

$$\begin{aligned}\frac{-1}{p-1}+\frac{n-r(X)}{\chi^2_{n-r(X)}\left(1-\frac{\alpha^*}{2}\right)}\cdot\frac{\hat{\sigma}^2\,[1+(p-1)\hat{\rho}]}{\sigma^2(p-1)} &\le \rho\\ &\le \frac{-1}{p-1}+\frac{n-r(X)}{\chi^2_{n-r(X)}\left(\frac{\alpha^*}{2}\right)}\cdot\frac{\hat{\sigma}^2\,[1+(p-1)\hat{\rho}]}{\sigma^2(p-1)}\,, \qquad (11)\end{aligned}$$

and inequalities (10) two increasing bounds:

$$1 - \frac{(p-1)(n-r(X))}{\chi^2_{(p-1)(n-r(X))}\left(\frac{\alpha^*}{2}\right)} \cdot \frac{\hat{\sigma}^2 (1-\hat{\rho})}{\sigma^2} \leq \rho$$
$$\leq 1 - \frac{(p-1)(n-r(X))}{\chi^2_{(p-1)(n-r(X))}\left(1-\frac{\alpha^*}{2}\right)} \cdot \frac{\hat{\sigma}^2 (1-\hat{\rho})}{\sigma^2}. \quad (12)$$

The region marked out by these bounds will be called pseudo-rectangular confidence region. It can also produce another conservative confidence interval for σ^2, which is given by the intersections of upper increasing with lower decreasing and lower increasing with upper decreasing boundaries:

$$\sigma_l^2 = \frac{1+(p-1)\hat{\rho}}{p} \cdot \frac{(n-r(X))\,\hat{\sigma}^2}{\chi^2_{n-r(X)}\left(1-\frac{\alpha^*}{2}\right)}$$
$$+ \frac{(p-1)(1-\hat{\rho})}{p} \cdot \frac{(p-1)(n-r(X))\,\hat{\sigma}^2}{\chi^2_{(p-1)(n-r(X))}\left(1-\frac{\alpha^*}{2}\right)}, \quad (13)$$

$$\sigma_u^2 = \frac{1+(p-1)\hat{\rho}}{p} \cdot \frac{(n-r(X))\,\hat{\sigma}^2}{\chi^2_{n-r(X)}\left(\frac{\alpha^*}{2}\right)}$$
$$+ \frac{(p-1)(1-\hat{\rho})}{p} \cdot \frac{(p-1)(n-r(X))\,\hat{\sigma}^2}{\chi^2_{(p-1)(n-r(X))}\left(\frac{\alpha^*}{2}\right)}. \quad (14)$$

It is worth noting, that the bounds are convex combinations of (6) and (7) using estimated value $\hat{\rho}$ and α replaced with α^*. Change from α to α^* does not enlarge the interval too much, as shown in the following figure, usually only by several percent. Figure 6 shows prolongation of the interval as a function of degrees of freedom for $\alpha = 0.05, 0.01, 0.001$ (from top to bottom). It follows that if ρ is substantially less than 1, the interval can be shorter than (6), especially for large p.

The other two intersection points of boundaries are given by the same linear combination of mutually interchanged terms.

Comparison of the two types of confidence regions (exact – dotted line and pseudo-rectangular – dashed line) for some choices of parameters and design is in Figures 7 – 9 for $\alpha = 0.05$.

We see that the approximate confidence region performs satisfactorily. Figures also contain α-cut (solid line) and α^*-cut (dash-dotted line) of the joint density $f\left(\hat{\sigma}^2, \hat{\rho}\right)$ for true values of parameters equal to the observed estimates. We can see that while α-cut produces regions with coverage

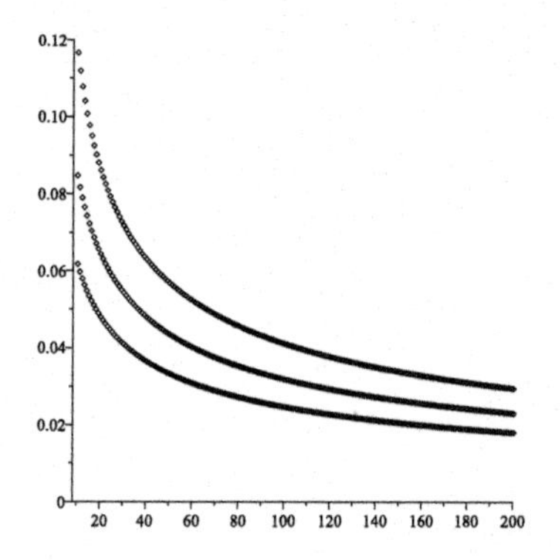

Fig. 6. Relative prolongation of the CI.

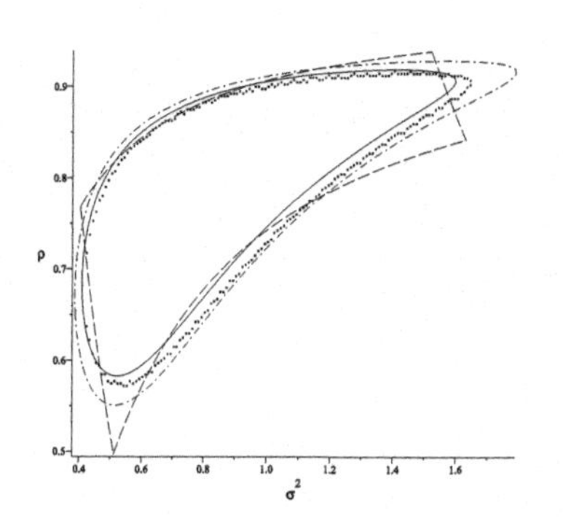

Fig. 7. $\hat{\sigma}^2 = 1$, $\hat{\rho} = 0.9$.

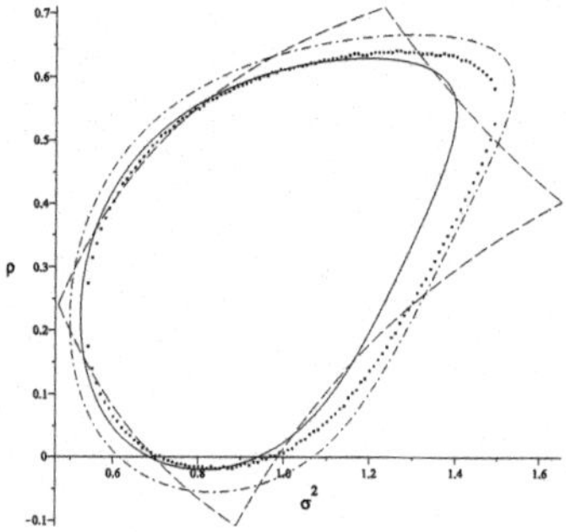

Fig. 8. $\hat{\sigma}^2 = 0.8$, $\hat{\rho} = 0.3$.

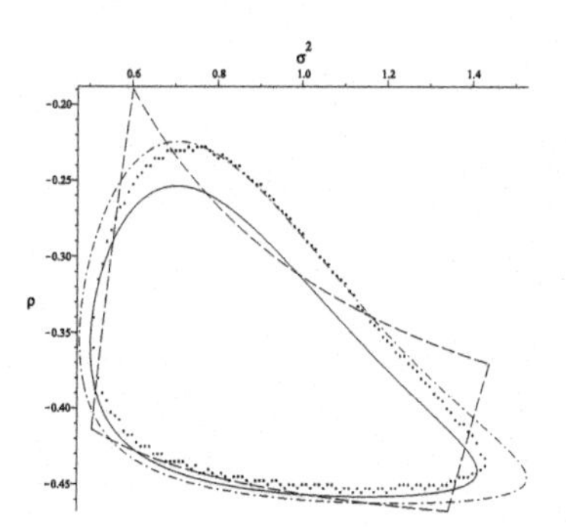

Fig. 9. $\hat{\sigma}^2 = 0.8$, $\hat{\rho} = -0.4$.

less than $1 - \alpha$, α^*-cut produces coverage greater than $1 - \alpha$, at least in all situations considered. It was confirmed by a limited number of other simulations for different α's. As a rule of thumb we can say that α^*-cut of the estimated density can be used as a conservative $(1 - \alpha)$-confidence region.

3. Conclusion

Depending on our needs, we can construct confidence regions for ρ and σ^2 independently, or simultaneously. Because of rather skew joint distribution of the estimators, we recommend to use the simultaneous confidence region approach. Exact (optimal) confidence region can be numerically computed. Pseudo-rectangular and conservative α^*-cut confidence regions, which are much easier to construct, can also be recommended for use.

Acknowledgement

The research was supported by grants VEGA MŠ SR 1/0131/09, VEGA MŠ SR 1/0325/10, VEGA MŠ SR 1/0410/11 and by the Agency of the Slovak Ministry of Education for the Structural Funds of the EU, under project ITMS:26220120007.

References

1. I. Žežula, *Statistics* **24**, 321 (1993).
2. J. C. Lee, *J. Amer. Statist. Assoc.* **83**, 432 (1988).
3. T. Kanda, *Ann. Inst. Statist. Math.* **4**, 519 (1992).
4. T. Kanda, *Hiroshima Math. J.* **24**, 135 (1994).
5. N. R. Chaganty, *J. Statist. Plann. Inference* **117**, 123 (2003).
6. I. Žežula, *J. Multivariate Anal.* **97**, 606 (2006).
7. Q.-G. Wu, *J. Statist. Plann. Inference* **69**, 101 (1998).
8. Q.-G. Wu, *J. Statist. Plann. Inference* **88**, 285 (2000).
9. D. Klein and I. Žežula, On uniform correlation structure, in *Proc. Mathematical Methods In Economics And Industry (MMEI 2007)*, (Herl'any, Slovakia, 2007).
10. R.-D. Ye and S.-G. Wang, *J. Statist. Plann. Inference* **139**, 2746 (2009).
11. I. Žežula and D. Klein, *Acta Comment. Univ. Tartu. Math.* **14**, 35 (2010).

AUTHOR INDEX